The Joy of being Eaten

The Joy of being Eaten

Journeys into the Bizarre Sexuality and Private Love Lives of the Ancient Layers of the Human Brain

LAWRENCE VANDERVERT

Copyright © 2014 by Lawrence Vandervert.

Library of Congress Control Number:		2014920596
ISBN:	Hardcover	978-1-4990-7454-3
	Softcover	978-1-4990-7455-0
	eBook	978-1-4990-7456-7

All rights reserved. No part of this book may be reproduced or transmitted in any form or by any means, electronic or mechanical, including photocopying, recording, or by any information storage and retrieval system, without permission in writing from the copyright owner.

Any people depicted in stock imagery provided by Thinkstock are models, and such images are being used for illustrative purposes only.
Certain stock imagery © Thinkstock.

This book was printed in the United States of America.

Rev. date: 01/28/2015

To order additional copies of this book, contact:
Xlibris
1-888-795-4274
www.Xlibris.com
Orders@Xlibris.com
671525

CONTENTS

Foreword 9
Prologue 13
Chapter 1 Naked Images from the Brain 15
Chapter 2 My Inner Body Is Black? 21
Chapter 3 A Thing of Beauty 26
Chapter 4 The Nano-soft Lecture:
           Dinosaurs Live in Our Brains 36
Chapter 5 The Walking Dead and Planet of the Apes 43
Chapter 6 Living Windows 56
Chapter 7 Flight to Eternity: Talking Boxes in Nonspace 72
Chapter 8 Two Romantic Dinners at Once 97
Chapter 9 The Journey 1: Into Nonspace and the
           Joy of Being Eaten 110
Chapter 10 The Journey 2: Caress Me—The New World Order 126
Afterword 167

Many Different Worlds, Many Different Titles

The Joy of Being Eaten offers new themes of science and science fiction that will be wholly new to people around the world. It is important to recognize that while people all over the world have similar interests and tastes in science and science fiction, the different themes running through *The Joy of Being Eaten* will have slightly different meanings and levels of appeal in different historical and cultural contexts. For this reason the title of this book emphasizes different themes in different countries. These "country-aligned" book titles include the following:

The Joy of Being Eaten (United States and Western Europe)

Living Windows (South Korea)

The All Time (India)

Brain-Mate (United Arab Emirates)

Reptilian Ecstasy (Japan)

In Full View (China and Indonesia)

JOURNEYS IN THE ALL TIME-2044 (Russia and Eastern Europe)

The Forever Time (United Kingdom)

The All Time (Australia)

The Answer (Tibet)

No matter what the title or the country, *The Joy of Being Eaten* reveals how the unconscious human mind came to be, and how we can soon journey back through those times that were both amazingly shocking and amazingly beautiful.

FOREWORD

Something Amazing is Happening in Science Fiction

In 1955, the famous brain scientist Wilder Penfield gave a report on how he was able to stimulate the brain so that people would remember past experiences as if they were happening again. He concluded that everything we have ever experienced is perfectly recorded deep in our brains. Then, in the late 1990's William Dobelle successfully put artificial vision directly into the brain of a blind patient. Amazingly, these researches show that our complete mental experience can be accessed by computer systems. Now, in 2015, brain researchers at the University of California in Berkeley are using MRI's to "read" visual images going on in the brain. See more about this new research that literally "reads minds" below and on the Internet at http://gallantlab.org/.

When added to the already existing advances in how the brain produces our holograms of experience and artificial intelligence (intelligent machines and androids), all three of the above startling scientific "mind-reading" advances are about to change the relationship between science and science fiction. Advances in science are now so stunning, they are catching up with the speed of the human imagination. Will they soon be running ahead of the human imagination?

*The Joy** builds directly on all of this fascinating new research to describe how even the "self" can come to life in a huge, 3D *Fullview* computer monitor. Thus, we can now read the complete human story that was recorded in our minds across millions of years. As we read the minds of two young women, we are taken back through the ancient history of the dinosaurs (which is the most ancient history of our minds), and of Homo erectus humans who lived 2 million years ago.

A New Level of Science Fiction is Emerging

There have been at least three major revolutions in science fiction: 1. Evolution (the latest and most interesting are the *Planet of the Apes* films); 2. Einstein stuff (the best was *Star Trek's* warp drive that took us beyond the speed of light); 3. Artificial Intelligence (the most fun were the droids R2-D2 and C-3PO in Star Wars).

2015: Reading Minds: An Emerging 4th Revolution in Science (and therefore a new revolution in science fiction)

Brain researcher Professor Jack Gallant and his team at the University of California at Berkeley have recently developed computer software that uses brain-imaging data to reconstruct actual images from the brain. See it at: http://gallantlab.org/.** I believe this technology will eventually allow us to see every detail of what people are thinking, to actually read minds! And then, with more development, to record the entirety of a person's thinking, personality, and feeling. What was once science fiction will be completely REAL! However, this wonderfully imaginative new research does not address a way to get the living, thinking individual (you) out of the brain and into the form of a non-biological computing system.

---

* *The Joy of Being Eaten* title is used in this forward only because this book was first published in the United States.

The 4th Revolution in Science Fiction: *The Joy of Being Eaten*[**]

However, there is a way to take the images that Professor Gallant is beginning to put the entire living, thinking person into a holographic electronic flux. This can enable you or anyone else to actually live in two systems (two realities) at once, one biological and one machine, or whichever one serves your long-rage purposes. *The Joy of Being Eaten* is the story of how that was first accomplished and the amazing journeys that await all of us deep inside the brain. In *The Joy of Being Eaten*, science and science fiction are exactly the same thing!

Lawrence Vandervert, Ph.D.

PS: If you are a student (or a professor), you will find that this novel enjoyably and intricately combines psychology, anthropology, paleoanthropology, computer science, artificial intelligence, the physics of Albert Einstein, how the brain produces holograms, and brain-imaging in ways that make you feel right at home.

Questions? Please visit me at my website: www.larryvandervert.com

Betty Jean Vandervert directed the lab where experiments leading to sections of this novel were conducted. Betty Jean also kindly edited the lab results for this book.

[**] Cited in this Forward with the kind permission of Dr. Jack Gallant, Professor of Psychology and Neuroscience, University of California at Berkeley.

PROLOGUE

San Francisco (Late, 2013)

As I stood on the deck of my yacht, the *Ti Amo*, I looked out across San Francisco Bay. The fog was dispersing, and it was beginning to rain. Two computer scientists who, like me, knew the eminent neuroscientist Karl Pribram were about to arrive at my slip at Pier 39. Pribram had shown the world how the brain makes holograms, the 3-D pictures we see, think, and dream about. The scientists were coming to meet *me*, the recently "famous" Colonel Edythe Peachey. They want to try to talk me into going to Nano-soft's headquarters up in Seattle. I won't go; I love having breakfast every morning on my *Ti Amo* too much! But I will let them talk to me—I also love having company.

The first drops of the morning rain pelted my hand, as I thought about how, a couple of decades ago, brain researchers in Utah had discovered that simply stimulating the surface of the visual areas of the brain caused people to see light patterns. I recalled the days back in 2009 when we had begun to put visual images directly into the brain. Our mission was to give sight to the blind. Following revolutionary artificial vision techniques pioneered by William Dobelle, we were gently pressing electrode arrays directly onto the delicate, gelatinous surface of the brains of several blind patients.

A 2009 picture I have on my desk in the *Ti Amo* shows the miniature video camera attached to the frames of a regular but lensless eye glasses that fed scenes we had set up in our lab to a small laptop computer. The computer then fed dotted patterns of electrical jolts representing the scenes directly to the array of electrodes pressed against the brain. Like dozens of tiny matches going off in the brain, these mild pulses of electricity gave light to brilliant phosphene pictures in the darkness of our blind patients.

Our patients could easily interpret these pictures, enabling them to read large letters and to navigate around objects we had set up in our lab. This was a mammoth advance from a world of darkness to a simple electrode pattern that would light up the world of sight for our patients! Our blind patients could see! What our patients saw through their miniature video cameras was also displayed for us in our large ball-shaped 3-D computer monitor we called *Fullview*.

Refinements in what our patients were able to see had come right along on a regular basis. One of our patients, who had lost his sight in an accident at age fourteen, told us that our electrode arrays, which contained 1,024 light points or pixels of moving scenes, were as good as a video version of an Etch A Sketch drawing toy. I remember his excitement; he was a typical boy and wanted to use his newfound vision to learn to create video games. Then through rapid advances in our ability to selectively microstimulate the brain, in early 2013, we were able to put television-quality "programming" into the brains of our patients. Of course, we were also observing these wonderful, high-quality scenes in Fullview as the patients moved their miniature video cameras through the lab during vision tests. Because we knew from Karl Pribram's work that the brain saw things as holograms, our next step was to put the lab scenes into Fullview in the form of actual holograms.

But the very speed of our progress was propelling us toward a huge snag. The television-quality pictures streaming into the brain from the electrode array was something totally new to the dark world of our patients and began to mix and compete with the patients' own ongoing thoughts. The patients reported that this sudden mixing of images was like awakening from a dream and not knowing what was real. Of course, these dreamlike, image-mixing experiences were extremely confusing and frightening for the patients. The image-mixing was so frightening, it threatened to shut down our research.

But, still, a few of the blind patients are mentally strong and willing. We continue to carry on with them at our Palo Alto lab.

CHAPTER 1

Naked Images from the Brain

(Palo Alto, January 14, 2014)

Jay Headley, our lab training coach, connected Pauline's brain electrodes to the miniaturized video camera mounted on her lensless eyeglass frames. Pauline, an attractive young blind patient, was smiling as always. Through the electrodes, the miniaturized camera sent scenes of the lab directly into Pauline's brain. Headley was a tall, thin thirty-three-year-old with a severely hooked nose and an attitude to go with it.

The scene showing the lab's testing area popped brightly onto the Fullview computer display. There in Fullview were the usual three chairs we would arrange as obstacle courses, two mannequins that our blind patients would learn to dress, and on the far wall, some standard tests for visual acuity. The scene in the Fullview monitor told us Pauline's camera was pulsing images into her brain and that she was seeing everything in the lab okay.

"Pauline, did you just mess around with your video camera?" Headley snapped with a slight smile. "You know I don't like patients doing that without telling me."

"No, I didn't do *anything*, Jay," Pauline replied. "I haven't moved even a tiny muscle."

"Huh, that's fun . . . ny," said Headley, his voice trailing off, as he peered intently into the Fullview monitor. "A shadow or something just moved across Fullview."

"Maybe it was a smudge on my camera lens or maybe just a fly," Pauline suggested.

"Impossible," Headley responded. "I meticulously clean these camera lenses before each session, and nobody's seen a bug in this lab for the last two years you've been coming here! Hey, Dr. Peachey, come over here a minute, would you?"

I had overheard Pauline and Jay's little commotion about the camera and was already on my way over to see what the problem was. As I approached, I caught a movement on the Fullview monitor out of the corner of my eye. The movement was very strange because Pauline was absolutely motionless. Normally, the monitor would only show the camera picking up how the blind patient was moving their hands while moving a chair around or putting a hat on one of the mannequins.

"Now what the hell was *that*?" said Headley, sounding a little startled and at the same time irritated.

Headley and I exchanged a quick, puzzled glance and became silently transfixed on the computer monitor.

"Headley, be a dear and go get Dr. Vandervoort—pronto!" I blurted out.

"I am on my way, Edythe," replied Headley as he shot up from his chair.

"Ah, never mind Headley, here he comes now," said Peachey.

Headley fell back into his chair, his eyes fixed on the monitor.

"What's going on?" asked Pauline. "Is that *my* hand in there? It can't be. I'm not doing anything."

I didn't know *what* was going on. A shadowy, hand-like object was moving here and there in the Fullview monitor. It seemed to be reaching and grasping toward one of the mannequins in Fullview's lab scene. As we watched, the hand emerged more fully. It seemed to be suspended on a small, spindly arm. Professor Vandervoort came into the lab and slipped into Headley's chair, gently pushing Headley out the other side. Vandervoort could do that; it was his way, and besides, he was the lab chief and the only neurologist on the staff.

"Professor, what the goddamn hell *is* that?" asked an obviously freaked-out Headley, who was peering over Vandervoort's shoulder.

Vandervoort sat staring at the arm and hand as they moved among the mannequins in Fullview. He didn't say a word for what seemed a full minute. Headley and I looked quizzically at each other several times.

"Pauline," Vandervoort said, finally breaking the silence, "what are you thinking about?"

"I don't know," Pauline replied. "I guess I was just thinking about picking the hat up from the floor and putting it on the small mannequin, and right now, I am getting ready to get up and actually do it."

"My god," Vandervoort quietly muttered, "I never would have believed it possible."

"Do you realize what that is, Edythe?" he added, not taking his eyes off the computer monitor.

"Is it some kind of feedback overlay or reflection?" I replied, shrugging my shoulders as I continued to watch the monitor.

Vandervoort continued to look at the hand moving about in the monitor as he stroked pensively across his neatly trimmed moustache and goatee.

"Come on, you guys, both of you know that each of us has a complete, detailed copy of our body inside of our brain—anybody who's taken Psychology 101 should know that," Vandervoort gently scolded us while smiling over at me. "It's the inner body that causes all of our movements, and it feels everything for us. When we move, it moves. When we run, it runs. When we are caressed, it feels pleasure. When we stub our toe, it feels the pain. When we have sex . . . well, you get the idea—it's having all the fun. It's our *real* body, where we actually do everything and really feel everything we do. It's a copy of our body laid out in nerve tracks inside our brain."

Coming out of the blue, Dr. Vandervoort's suggestion of an "inner body" sounded bizarre. But then I slowly began to recall the motor-sensory homunculus from psychology classes that I had taken many years ago at Stanford.

"Yes," I said, quickly picking up on Professor Vandervoort's impromptu little lesson about the inner body. "I remember my psychology professor at Stanford ran around his podium, waving his arms and telling us that a duplicate body existed inside our brains and that it felt and did everything for us. Even when we dream, he told us, it's the 'you' that does and feels everything. And that this inner body is the only reason you are able to do things in dreams even though our body is motionless! And then he said something I will never forget. He told us, 'In fact, it's the real "you," the real you that's looking out through your eyeballs!' That was maybe the most memorable lecture from my university days—the real 'me' was actually separated from my external body—it was an impulse pattern living in the circuits of my brain!"

Little did I know at the time the incredible truths that this fact from my days at Stanford would reveal in the months to come.

"So, Professor Vandervoort, you guys are saying that that's Pauline's *own* arm and hand moving around in the Fullview monitor?" Headley asked in a smart-aleck but nervous disbelief. "That's Pauline's 'inner' hand in there?"

"Yes, that's the hand of Pauline's motor-sensory homunculus," I interjected. "That *is* what you're saying, isn't it, Hans?"

"Yep," replied Vandervoort. "That's exactly what I am saying!"

"Pauline," Vandervoort said, "would you please imagine reaching out with both hands and picking up the hat and putting in on the little boy mannequin?"

Two huge hands on spindly little arms thrust themselves forward in Fullview; the back of a large head with small shoulders followed them into view.

"Holy Christ," yelled Headley, "will you look at that!"

Everyone went silent. We could not believe what we were seeing.

"Why is Pauline's homunculus suddenly starting to appear on the monitor on its own, so to speak?" I finally asked as I walked around to check out the lens of Pauline's video camera.

"I think Pauline's homunculus is beginning to show up for exactly the same reasons we were shutting down this research program," Vandervoort replied.

"What do you mean?" I asked, still somewhat taken aback by what I was seeing.

"Edythe, our whole problem was that our blind patients were becoming confused about what was real," Vandervoort explained. "Remember how our patients couldn't tell if they were experiencing their own thoughts or the video images coming directly into their brains through their electrode arrays?"

"Yes," I cautiously replied, not quite sure where Vandervoort was going.

"I think the blind person's equally blind homunculus had the same problem," Vandervoort went on. "Let me explain. We all have two picture shows going on in our heads at the same time, one coming from our own thoughts and one coming in through our eyes. Since the homunculus of the blind patient is not familiar with the visual world, the homunculus began to become confused when we put television-quality picture shows in through the electrode arrays. I should have realized this. Living in the world of the blind person, the homunculus had only experienced its own ongoing thoughts—it had only experienced a single picture show. When we began putting television scenes into the electrode arrays in the brain, it was just too confusing for the homunculus. The two picture shows began to mix together, and the patient's homunculus suddenly didn't know whether it was thinking or dreaming. It suddenly didn't know what was real. So out of fright and desperation, the homunculus, the inner person, automatically began to migrate toward the electrode to see where the confusing pictures were coming from."

"Wow," exclaimed Headley. "That's damn weird—damn weird!"

"Ah, yes, I think I see what you are saying, Hans," I chimed in. "That's why it showed up in Fullview! God, how simple. It's the same sort of thing that happens when a sighted person is in a dark room and sees a strange light or hears a strange sound—it piques their curiosity, and they have to go see where it's coming from."

"That's right!" replied Vandervoort. "And when Pauline's homunculus finally found the electrode array and began to touch and feel it, through that same touching, it simultaneously migrated—or 'crawled,' you could say—into the Fullview monitor, where we could see it!"

The appearance of Pauline's hand in the computer monitor had caught me off guard. It had taken a few minutes for the significance of what we were seeing to sink in.

"This is absolutely astounding!" I exclaimed. "Do you realize we are actually seeing mental images moving around inside Pauline's brain? Pauline only imagined reaching out her hands, and there they are in Fullview for everybody to see! For the first time in history, we are seeing part of another person's private world of . . . of thoughts! At least I think they're thoughts."

"Yes, Edythe," replied Vandervoort, "but more accurately, we are seeing the raw, deeply naked Pauline carrying out her own thoughts. But you're right, we *are* seeing Pauline's thoughts. Whatever we see Pauline's homunculus doing, that's what she is thinking about. We can see what Pauline is thinking!"

The lab was completely silent. Everyone just stared into the Fullview monitor.

"And seeing Pauline's inner body coming to life so vividly 'outside of her body' reminds me of something else," Vandervoort said, finally breaking the silence. "It reminds of a very strange condition that only rarely pops up in books on psychiatry."

CHAPTER 2

My Inner Body Is Black?

(Palo Alto, 2014)

"Psychiatry?" I asked with a puzzled tone. "Oh, oh, yes! The psychiatric reports of body doubles! What a great mental leap, Hans, I think I know where you are going. What we are seeing in Fullview is sort of a doppelgänger, isn't it? Is that what you're thinking?"

"Yes, Edythe," replied Vandervoort. "Pauline's homunculus was forced to migrate to her electrode array and come to life in the Fullview monitor when she was deeply confused. In the same way, doppelgängers sometimes appear before the eyes of people who have become very deeply confused."

"Doppel-what-ers?" asked a totally bemused Headley.

"Doppelgängers are body image *doubles* that appear to some people to be outside of their body," I replied. "It's a German word that means 'double walker.' People have experienced them on and off down through history. Some have seen their body doubles walking in front of them, standing beside them, and have even felt them trying to do things for them."

"Yes, that's right, Headley," Vandervoort replied, slowly nodding his head in agreement as he continued to peer into the monitor. "No one has been able to figure out where doppelgängers come from or exactly why they appear. This thing, this inner body we are seeing in the Fullview monitor could be the answer. It could be the prime source or neural matrix for the doppelgänger that everyone has waiting inside their brain but hardly anyone ever sees."

"So you're saying the homunculus inside our brain is the real, living 'us' that does all of our feeling and thinking, as Dr. Peachey said," replied a slightly incredulous Headley. "And it is so real that under rare circumstances, it can even take on a sort of life of its own?"

"Yep, strange as it may sound, that's precisely what I am saying," said Vandervoort as he glanced over at me. "Think about it. Just as Dr. Peachey's psychology professor said, our outer body is really a puppet body that jumps around, doing things only because the homunculus inner body jumps around in the brain, telling it to. But the inner body is also the *real* you that has out-of-body experiences and imagines hovering above the operating table when people have near-death experiences. It is this real inner body that has phantom limb experiences in which it continues to feel a leg or arm even though it has been amputated. It is this real inner body that, instead of speaking to another person, 'speaks' to the schizophrenic's speech-input area and thus hears voices."

"That's nutty!" exclaimed Headley.

"I know you said Pauline's homunculus began touching and feeling the electrode array, Professor Vandervoort, but how exactly did Pauline's homunculus or doppelgänger manage to get out of her brain and into Fullview?" Headley went on. "I mean, how on earth did it actually get itself through the wires and *out*?"

"Good question, but think about it for a minute, Headley," Vandervoort answered in his serious professorial manner. "Pauline's homunculus began interacting with the thin 'sheet' of phosphene light patterns right at the surface of Pauline's electrode array. And because the homunculus manipulated the phosphene images being produced on the surface of the electrode array, the sheet becomes a 'mirror.' Everything Pauline's homunculus does, sees, and feels changes the electrical patterns on the surface of the array. This creates a 3-D, ten-thousand-pixel image on the electrode array that is identical to the homunculus. In other words, it creates a doppelgänger right on the surface of the electrode array!"

"Hey," announced Headley as he turned to Vandervoort, "that mirror image sounds like electrical impedance tomography!"

"What on earth is 'electrical impedance tomography'?" Pauline interrupted before Vandervoort could respond.

"Uh, uh . . . it's just a series of pictures showing the pattern of electricity dancing around your electrode," Headley answered with some hesitation. "As the electricity tries to get through the electrode, your homunculus gets in the way, and its silhouette produces an exact picture of what your inner body is doing."

"That's pretty good, Headley," Vandervoort added. "Pauline, electrical impedance tomography is being used more and more in brain studies these days. It's a little like MRI or even like ultrasound imaging in that it gives us a look at what's going on deep inside. Electrical impedance tomography gives us a picture of what's going on inside the brain millisecond by millisecond."

"But I am still wondering how the tomography or 3-D depth of Pauline's homunculus is being created at the electrode surface," Headley went on, again turning to Vandervoort.

"Good point, Headley," Vandervoort replied. "The homunculus is creating its own 3-D pictures by constantly moving around on the phosphene sheet as it does things. Millisecond by millisecond, every movement it makes changes the perspective of the picture at the surface of the electrode array. This gives us the 3-D view of every movement of Pauline's homunculus."

"Absolutely beautiful!" Headley nodded in agreement.

"Let me jump in here," I said excitedly. "This is a major, major breakthrough—we could win the Nobel for this. Let me get this all straight for my report on this! Pauline's computer is sending pictures into her electrode array. The electrode array is causing phosphene pictures to light up in her brain. Her homunculus gets on the electrode and interacts with those pictures, and that makes both her homunculus and the pictures show up together on our computer monitors, right?"

There was dead silence for a moment.

"Right, Dr. Peachey!" Vandervoort and Headley simultaneously exclaimed.

"Thank god for the 3-D impedance of my homunculus, whatever that is," Pauline quietly sighed.

"Let's put it this way, Pauline," Vandervoort replied. "3-D electrical impedance has miraculously allowed your homunculus to slip out of your brain and into Fullview. And now we can see what you are doing inside your brain. That means, my dear, we can see what you are thinking!"

"You, all of you, have my permission to watch me think," Pauline replied with a smile. "Although, Jay, at times, I will ask you to cover your eyes!"

With those simple words, Pauline had made us realize, I think, the depth of what we had stumbled onto.

"How do you *feel*, Pauline?" I asked. "I mean, do you have any feelings of confusion as you see your homunculus emerging into the Fullview monitor?"

"No, it's just weird," Pauline replied. "I can see my homunculus doing what I am thinking. And it feels like my homunculus is seeing and doing things for me. It's like it provides an extra set of hands and eyes so I can watch myself doing things and even watch myself thinking in Jay's Fullview monitor. My thoughts are not confused at all with what the homunculus is seeing. I guess my inner body has accepted the visual images coming in on the electrode array to be a real, outside world, and at the same time, my other thoughts are separate from it."

"Yes," said Vandervoort, "your homunculus can now manipulate what it sees coming in on the electrode array, or it can switch, so to speak, and pay attention to its normal thoughts as it always has. But this new learning is no different from what every sighted person goes through during infancy. Because sighted people can see while they are babies, they learn that what they see in the outside world and what they are thinking are clearly two different things. Now, you have learned to separate your own thoughts from the pictures coming in from the electrode array."

The vast implications of what we had been discovering were slowly seeping into my head.

"Correct me if I'm wrong, Dr. Vandervoort," I interrupted, "but this means we are going to be able to watch Pauline's thoughts, no matter what she is thinking. That could get pretty interesting, couldn't it?"

"It's a whole new world," Vandervoort thoughtfully replied, nodding his head. "Yes, Edythe, when Pauline's homunculus solved its confusion problem, it dropped a huge scientific breakthrough right into our lab. I have a feeling that that homunculus is going to be able to take us into some very secret and very strange places in Pauline's brain."

In Pauline's case—and we assumed the other patients would follow suit—the confusion that had been caused by the mixing of images had completely subsided. The discovery that a patient's homunculus would seek out and migrate to the surface of the electrode array completely changed our research plans.

Professor Vandervoort immediately redirected our research toward a more detailed and complete study of Pauline's homunculus. Since the images we see inside our brains are holograms, we quickly replaced our regular electrical-impedance-type electrodes with holographic, image-gathering electrodes. To say the least, the discovery of a way to view an actual "creature" inside the mind that could feel, hear, move, see, and think was, as Vandervoort said, an amazing breakthrough—more than amazing!

CHAPTER 3

A Thing of Beauty

(Palo Alto, 2014)

In order to study Pauline's homunculus, now materializing in more and more detail, Vandervoort had Headley bring in an experimental version of a larger 3-D computer monitor. The new monitor looked like a huge, elongated crystal ball. It was a little larger than a Volkswagen Beetle. Since we would be able to view this new monitor from 360 degrees, Professor Vandervoort had chairs arranged all around it. This state-of-the-art monitor was under development for the Defense Advanced Research Projects Agency (DARPA) for the Airborne Warning and Control System (AWACS). DARPA had developed an algorithm that turned radar blips into electronic images that looked exactly like the particular airplanes they represented. AWACS air traffic controllers could now see perfect little copies of all airplanes within three hundred miles in one of these Volkswagen-sized glass monitors. The first time I saw the elongated "ball," as we first called it, I found two technicians working on it from the inside.

"Hi," I shouted as I walked up to the ball and knocked on its flawless, shimmering glass surface.

"Hi to you too!" said one of the people inside the ball as he turned to look at me. "You don't need to shout," he went on. "We can hear a pin drop down there where you are in Palo Alto."

"What do you mean 'down there' where I am in Palo Alto?" I replied, a little perplexed at such a peculiar comment. "You're *here* in Palo Alto too, aren't you?"

"No, no, we are in Redmond, Washington, at Nano-soft," the female technician inside the ball replied.

Headley walked up behind me. "Aren't these virtual-experience computers something?" he said. "Hey, you guys," he said, obviously familiar with the people inside the ball. "Dr. Peachey, meet Jim and Amanda."

"How can we just be talking back and forth with real people with real bodies clear up in Redmond like they are physically present here in Palo Alto?" I said in astonishment. "I mean, it is solid glass between us, they look totally real, and we're talking back and forth like it's an open window. And what are they doing *inside* the ball?"

"It does look like ordinary glass, Dr. Peachey," Headley replied as he ran his hand lovingly and delicately over the smooth, glistening surface of the ball. "But it's way, waaay more. It's a transparent two-way tubular retina called the Samsing Seamless. This baby is a high-def tube-shaped computer monitor both on the outside and the inside, and it has absolutely no seams, none anywhere. It's called a two-way tubular retina because it 'looks' inward and outward at the same time. The AWACS people have a huge version of the Samsing Seamless running the entire inside length of one of their research planes. The recent history on this thing is that DARPA bought the technology from Samsing for its AWACS planes, and that's pretty much all I know about it. We got this prototype just before DARPA went underground with its further development."

I didn't let Headley know how awestruck I was at the idea of a computer monitor that was a two-way retina. I knew the retina of the eye was made up of the cells that took in light and began the process of vision, but I didn't know an artificial retina could be made that looked both outward and inward at the same time.

"Why did Professor Vandervoort want such an 'exotic,' *two-way* computer monitor just to study Pauline's homunculus?" I asked Headley.

"Funny, I asked him that same question," responded a pensive Headley. "He just said, 'I want to be able to talk *directly* with Pauline's homunculus,' and then he laughed like he was joking."

I continued to wonder why Dr. Vandervoort would want a super-advanced two-way retina computer monitor. I wasn't as sure as Headley that Vandervoort was only joking.

"Are Jim and Amanda actually inside the ball?" I asked Headley. "They look like they are."

"They might be, might not be," replied Headley. "They're transmitting from a two-way retina themselves, so could be either outside or inside. You'd have to ask them which."

"Hey, Jim and Amanda," I said, turning to the ball. "Are you two inside your monitor or outside?"

"Hi, again, Dr. Peachey," replied Amanda. "Right now, Jim is outside, and I am inside. This tubular two-way monitor is open at one end, and we just walk in and out whenever we wish."

"Cool!" is all I could say.

"Thanks, Amanda," said Headley. "Bye, for now."

This is absolutely incredible! I thought. The possibilities of the tubular two-way retina for research with Pauline were staggering! Now I was becoming convinced Professor Vandervoort *wasn't* joking about somehow talking directly to Pauline's homunculus.

The next morning, Pauline was in the lab for her first session with the new ball monitor. Her holographic, image-gathering electrodes were attached. Pauline's video camera came on, and a perfect 3-D replica of the lab immediately appeared inside the ball. We finally got a super clear holographic view of the world that Pauline had been seeing. There seemed to be no separation between us and Pauline's world inside the monitor—the glass was so clear and the scene so vivid it seemed as if the part of the lab she was viewing had magically materialized before us out of nothing! So there we all were in the lab with, at the same time, a perfect duplicate of the lab suspended before us. Pauline's brain was looking at the lab, and in the Samsing, we were looking inside Pauline's brain! We had entered a new kind of world! Since we were sitting in a circle around the ball, each of us got a slightly different view right through Pauline's brain and into the lab.

"Pauline," Dr. Vandervoort said, "please imagine walking into the lab just as you did before with the old Fullview computer monitor."

Instantly, the tips of Pauline's huge homunculus fingers began to nucleate from "nothing" before our eyes.

"Here that baby comes," announced Headley, who was sitting directly on the other side of the ball from Pauline.

"Yes, I can see it now too," I added.

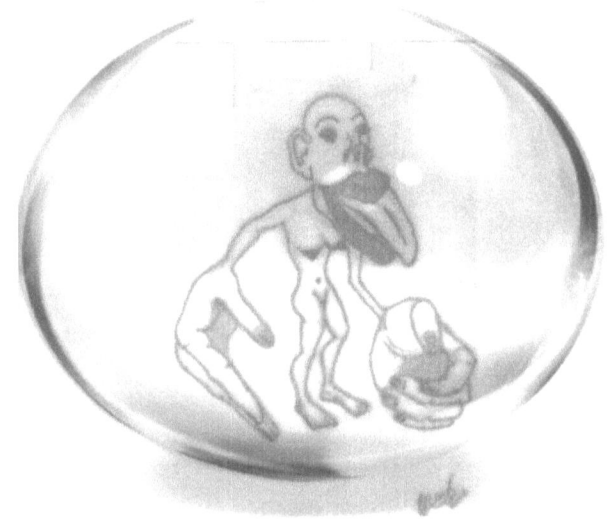

Vandervoort sat in silence, pensively stroking his moustache as he peered intently at Pauline's homunculus.

"Boy, she sure ain't shadowy anymore!" said Headley. "Wow, what a . . . an 'animal' that thing is! This two-way retina has made everything so clear and soooo 3-D! And it looks so 'touchable'!"

Headley was a little overdone, but he was right. What an animal it was! It no longer seemed we were sitting around a huge computer monitor; it seemed like we were in the front row of a weird, almost macabre live play! We were dumbstruck!

"All of the mental images in the brain, including the homunculus, are actually holograms," Vandervoort suddenly said so everyone could hear. "For the two-way retina, we have replaced the old electrode arrays with *holographic, image-gathering electrodes*. Before we got the ball, we saw the homunculus projected in 3-D but only onto a 2-D computer screen. Now, we will see Pauline's homunculus as it really is. We will see it as a hologram. We will see as we see the real world around us."

"So Pauline's homunculus is only a hologram?" Headley asked.

"I wouldn't say *only* a hologram, Jay," replied Vandervoort, again in a voice everyone could hear. "What you see in front of you is far *more* 'sophisticated' than any optics-made hologram you have ever seen. The hologram of Pauline's homunculus you see is a *living* thing—it's a living hologram. All of the holograms that your brain produces in your own head are 'alive.' They are part of your living brain. Pauline's homunculus is no different."

"That's right, Jay," I added. "Let me give you an example. Right now, as you look at me, the image of me that you are seeing *inside* your brain is a living hologram. In a very real way, I am alive in your brain!"

"Think of the possibilities," Vandervoort said, looking over to Headley with a wink and smile.

"Yeah, think of the possibilities," replied Headley as he turned his head away with a friendly grimace.

Our discussion moved on swiftly, no one at the time realizing the incredible implications of Professor Vandervoort's comment about the homunculus being somehow "alive." During that first day with Pauline's homunculus in beautiful, "touchable 3-D," we examined its intriguing body in full detail. We were able to determine more about the relationship between "Pauline" and her homunculus. While Pauline's thoughts and her homunculus continued to be perfectly coordinated most of the time, the homunculus would sometimes behave in puzzling ways. This "autonomous" behavior of the homunculus was very perplexing even to Professor Vandervoort, and for the time being, its explanation had to be left until we knew more.

We were stunned by the overall appearance of Pauline's homunculus. It was a bizarre body, yet it was somehow a thing of beauty as it moved gracefully around inside the ball. I now realized that Pauline's homunculus could see us because we and the ball were in the lab and were therefore part of what was being fed to Pauline's image-gathering electrode. Professor Vandervoort was obviously way ahead of Headley and me on this; he knew Pauline's homunculus would be able to see us with the two-way retina Samsing monitor. Because Pauline's mind was completely open to us and was displayed in full, *living holographic view* for everyone to see, we decided to transfer the name *Fullview* from our old computer monitor to the new ball-shaped Samsing Seamless monitor. Now it was *really Fullview*!

"Why is Pauline's homunculus so deformed, Professor Vandervoort?" Headley asked. "Well, I guess it's not so much deformed as it is just 'different.' Is it partly because Pauline is blind?"

Pauline raised an eyebrow and quickly turned her head to train her tiny video camera on Professor Vandervoort. Her homunculus simultaneously turned and looked at the internal surface of Fullview. It was looking right about where Vandervoort was sitting on the other side of the two-way retina. Along with Pauline, Pauline's homunculus seemed to be waiting for an answer.

"I see your homunculus is trying to look at me, Pauline," Vandervoort said with a grin. "Jay, Pauline's homunculus is perfectly normal. It looks pretty much like the homunculus in *your* brain. Remember, the sizes of the different parts of the homunculus's body simply depend on how much of the brain's cortex is devoted to that part of the body."

Pauline remained deafeningly silent. She was seeing herself for the first time since the age of two, when she became blind, and we could tell she was comparing the strange humanoid "being" she saw in Fullview with the rest of us in the lab.

"Good god, just look at the size of that tongue!" said Headley with his eyes nearly popping out. "It must be eighteen inches long and six inches thick."

I remembered from my psychology class at Stanford that the homunculi we were shown were all "males." Pauline's homunculus was clearly—very clearly—a female.

"Dr. Vandervoort," I queried, "I see that Pauline's homunculus is quite definitely a female." "When I was in university at Stanford, all they showed us was a male homunculus. Why didn't they show us a female version?"

"In those ancient days when you were at university, Dr. Peachey, they hadn't yet established the details of the female homunculus," said Vandervoort, smiling and glancing over to catch Peachey's reaction to the "those ancient days" comment. "Brain imaging studies just last year proved that the female homunculus is just as you see it in Pauline. All the details you see in Pauline's homunculus are part of how she moves about in the world, what she does with it, how she feels it, and how she thinks about it."

"It's got nipples!" Headley blurted out in surprise, hitting his forehead with the palm of his hand. "Mama Mia!"

"There, that's total confirmation for the imaging studies I just mentioned," noted Vandervoort. "By the way, Headley, it's not an *it*, it's a *she*."

"And look at my beautiful huge hands!" Pauline suddenly exclaimed jokingly but with pride and finally coming to life. "They look powerful yet skillful enough to play a piano or sculpt. They must be big enough to wrap around a refrigerator!"

"Look at your feet, Pauline," I laughingly added. "You wouldn't need any snowshoes with those babies!"

"The tongue, the hands, and the feet are oversized inside our brains for a reason," I commented, following up on what Vandervoort had said. "We humans use our hands, feet, mouths, and tongues more than anything else. We talk constantly, and we manipulate things constantly. It takes a lot of brain area to do those things."

"Her skin is sure a strange color—almost black," Pauline commented. "Am I black? No one has ever mentioned that I am black. And I don't remember anything like that from when I was a little girl, and I could see."

"No, Pauline, you're not black," replied Dr. Vandervoort.

"Then is it black because it lives in my brain, Dr. Vandervoort?" Pauline quickly followed.

"Pauline, that's an interesting question," replied Vandervoort. "I'm not exactly sure why it's that dark color. We do know, however, that the human-type homunculus evolved into existence before early humans began migrating out of Africa hundreds of thousands of years ago. That could be the connection. If this is true, then the skin representation in the brain of all of our homunculi is of Negroid origin. It's interesting that even though the skin coloration of the races has evolved since then, the skin representation *inside* the brain has not. I suppose it's another one of those vestigial characteristics like the appendix or goose bumps that have been left over from atavistic times."

"Hey, it just occurred to me," I added. "From a psychological point of view, the homunculus is what we all *really* look like. I mean, think about it. We have this blackish creature running around in our brains that sees, feels, hears, and does everything. Yes, it's a *mental* version of us, and that's just the point. It's what we really are. We humans are 'mental' beings. But at the same time, the homunculus has no idea what its own body looks like. It has never seen itself the way Pauline is now seeing 'herself' in Fullview. So our homunculi think they look like our *outer* body, the one they see in the mirror. The real us inside is fooled by the way we look in the mirror all our lives."

"So because she became blind at age two, too young to remember her own body, it's almost like Pauline has never really seen her own body?" Headley chimed in. "That's weird."

"But today, I have seen my own homunculus," Pauline piped up. "And so now I know what I really look like. I think she's beautiful. I'm glad it's me."

The lab was silent for a moment.

"And I see that girl wearing the glasses with the tiny camera and the electrodes," Pauline went on, training her video camera on our other monitor, which provided a view of everything going on in the lab.

"I think she's beautiful too!" Pauline went on after a short, thoughtful pause.

"Oh, for fu—" Headley started to rant.

"Jay, I'm glad you caught yourself," Vandervoort said in a reprimanding tone. "There's a sixteen-year-old girl in this room, and you *don't* want to mess with her mother!"

"Jesus, Professor!" Headley exclaimed. "All of this psychology stuff is getting too damn deep and weird for me!"

"Thanks, Hans," I leaned over and whispered in Dr. Vandervoort's ear. "I know Headley's the best in the business, but he gets a little squirrelly sometimes. Just between you and me, I kind of think he went on that little rant because he was actually touched by Pauline's revelations about herself."

Soon after our discovery that Pauline's homunculus had migrated to the sheet of phosphene activity on her electrode array, the news had been leaking out. But the *Seattle Post-Intelligencer* hadn't gotten the story straight. The rumor was that we had managed to implant a new type of computer operating system in the brain. Every advanced computer software lab in the country wanted to know about what the Seattle newspaper's science writer had dubbed the "Brain-Organizing Software System" (BOSS). The rumored idea was that with the BOSS implanted, people wouldn't need computers anymore because their brains would *be* computers. It's funny when I think about it in retrospect—the rumor was completely backward. Instead of an operating system going into the brain, the brain's operating software system, the real BOSS, the living homunculus, was actually on its way "out" of the brain.

Within three months, our developing "living homunculus technology" was sold into the hands of software designers at a Skunk Works outfit in Redmond, Washington, where we suddenly were under the joint funding of Nano-soft and DARPA. Professor Vandervoort, Jay Headley, and I were lucky enough to be able to stay with the project when it went to the new owners. At the Skunk Works, we began by using the homunculus technology to explore entirely new kinds of options for computer and Internet operating systems. Among insiders, the project was known as *Living Windows 2016*. *Living Windows* was Nano-soft's choice of a name for a whole new kind of 3-D computer windows. The operating system of windows was to be alive just like Pauline's homunculus.

Nonspace

By early May of 2014, our holographic image-gathering electrode was being used in probes that went deeper into the brain. The deeper probes allowed us to extract living images from more ancient parts of Pauline's brain. These images were of ancient homunculi, and their holographic memories were transmitted into steadily larger, more advanced ball-shaped 3-D monitors. We had become used to the fact that although Pauline might just be sitting in her chair, "thinking," as she waited for a research session to start, we would see her homunculus begin to carry out a strange variety of activities on its own. We, of course, knew that this was happening as a result of Pauline's homunculus migrating to the electrode array on the surface of her brain. Professor Vandervoort had told us that whenever people think or imagine *anything*, their homunculus can carry it out anywhere within the holographic reality of the brain. He explained that when we imagine or think about going to the grocery store, flying to Mars, solving quadratic equations on an algebra test or even when we dream, our homunculus is actually going through imaginary motions in the holographic "nonspace" inside our brain. The entire inside of the brain is a playground for the imagination of the homunculus.

To scientifically check Professor Vandervoort's whole story about why Pauline's homunculus was moving about, doing strange things in the 3-D computer monitor even when she was only imagining, we hooked Pauline up to our electroencephalograph and had her imagine flying to Mars. To our amazement, Pauline's homunculus in the 3-D Fullview monitor took on the position of flying like Superman, and we were getting electrical activity from all over her brain. Stars from holographic memory appeared all around Pauline's homunculus in the monitor, and within a few seconds, an image of the planet Mars began to appear with all its classic canals and its polar ice caps. We were stupefied. Pauline's homunculus had summoned holographic memory images stored in her brain of what the blind girl had learned about Mars. She had traveled to Mars not in the real space of a one hundred million miles and the real time of several months but in her holographic, nonspace reality. This nonspace reality was free of any particular, identifiable space and the regular passage of time. Yet we and Pauline could see it. She could imagine it happening in the nonspace of her own brain, and through the holographic image-gathering electrode, we could see it happening in the nonspace of the 3-D Fullview computer monitor. Through her homunculus, we had brought Pauline's imagination right out of her brain and into our world of the 3-D Fullview computer monitor. Since when Pauline was thinking, her homunculus didn't ever really go anywhere, from then on we simply called the whole imaginary world, both in the brain and as it is seen in Fullview, *nonspace*.

CHAPTER 4

The Nano-soft Lecture: Dinosaurs Live in Our Brains

(Redmond, Washington, June 12, 2014)

It was mid-June 2014. I was a psychology graduate student with a bachelor's degree in computer science attending a special faculty conference on the Nano-soft campus in Redmond, Washington. It was the morning of the second day of the conference, and we were in breakout sessions, listening to the latest research from leading information scientists.

I had chosen to attend a session by a retired researcher by the name of Dr. Edythe Peachey. She had told us about her lab's experiments that put visual images directly into the brains of blind patients. Then in the last part of the session, she created quite an uproar when she showed the thirty or so of us a video clip of a weirdly shaped creature she claimed was the actual motor-sensory homunculus that came right out of the brain of a young blind patient named Pauline. In the video, this strangely shaped but oddly attractive creature was shown walking around in a ball-shaped 3-D computer monitor called Fullview. Its lips, tongue, mouth, and hands were huge! At the end of the video, we all sat in dumbstruck silence.

"Oh, come on, Dr. Peachey, that can't really be coming from Pauline's brain," a researcher suddenly shouted from the back of the conference room. "You're just trying to see how gullible we are. You had your graphics people cook up that homunculus, didn't you? I mean, the homunculus was even a strange dark color."

According to Dr. Peachey, this distorted replica of Pauline's body was no trick and was indeed real. She told us that it had apparently migrated onto the holographic image-gathering electrode that had been permanently attached directly to Pauline's brain. She told us we were actually seeing mental images from inside Pauline's brain!

Wow, I thought, *I wonder what it would be like if we could see what Pauline's homunculus was doing all the time, not just when she was putting a hat on a mannequin as we saw in the video. What would it be like to see what the homunculus was doing when Pauline thought about other things, like when she thought about her boyfriend or about her childhood?*

"Does anyone have further comments or questions about Pauline's homunculus?" Peachey asked as she slowly gazed around the room, her eyes curiously finally resting on me.

Only because Peachey happened to be looking at me, I timidly raised my hand about halfway.

"What is your name?" she asked as she cocked up one eyebrow.

"Betty Jean, or just BJ's okay," I blurted out, wondering why she had chosen me as her channel to the rest of the audience.

"I think it is amazing that we can see mental images from Pauline's brain," I said in a quiet, controlled tone that was totally out of character for me.

"But it seems even more amazing that the images are of Pauline doing things inside her own mind," I went on. "I mean, god, we can see what's going on in her brain! When her homunculus put the hat on the mannequin, it's like we could secretly see what Pauline was thinking. Could a holographic, image-gathering electrode be permanently implanted in the brain to somehow allow us to see what the homunculus was doing all the time? Could we use the homunculus to see what people are thinking? It would be great to use something like that in a police interrogation or, for that matter, in any interview. What would it be like to see what other people are thinking?"

"Whoa!" responded Peachey. "My, Betty Jean, you have so many questions."

Dr. Peachey explained to us that most psychologists would probably say that seeing other people's thoughts would be out of the question. They would say the possibility of seeing thoughts would be ridiculous, laughable, and completely absurd—simply out of the question.

"But," she quickly added, "to hell with most psychologists! Ninety-nine percent of them don't know what they're talking about anyway, and those in the other 1 percent are like me—they're weird misfits."

"So, Professor, you're a weird misfit," one of the other scientists interrupted with a grin, "and maybe you don't know what you're talking about?"

"Touché," Peachey responded with an appreciative smile. "Yes, I'm weird. Let me tell you a little story about how weird I really am.

"I spent most of my childhood in a little town just outside of Seattle," she went on as she walked toward one of the windows overlooking the Nano-soft campus.

"Our local grocery store had cow brains in their butcher shop," she went on, glancing over her shoulder at us, stopping, and again raising one eyebrow.

"Cow brains are a little out of style these days, but you can still find them. Some butchers will special order them for you," she continued.

God, she is really weird, I thought to myself.

"Every time my mother and I would go to the store," Peachey went on, "I was mysteriously drawn to the cow brains. My mother must have wondered what she was raising. The brains were all neatly packed and pressed up against the bulging cellophane. The spiderweb channels of blood exposed every convolution. I would stare at those brains, wondering what the cow had seen during its life and if there were still memories lurking in there. Cow brains were different from regular meat. I couldn't help thinking how these packages of 'meat' saw, heard, and smelled the world and carried memories of Farmer Brown coming and going about the barn and pasture. The traces of a life of experiences, habits, pain, pleasure—all those things waited to be purchased for twenty-nine cents a pound and then taken home and eaten for breakfast! How strange, I thought, eating another organism's experiences of life. The whole situation seemed impossible. It haunted me deeply. Every time I saw those brains, I vowed that someday I would figure out how to look into that brain and see the cow's world. I knew it had to be possible."

As she talked, Peachey had continued to stare out the window as though she had forgotten we were there. Suddenly, she turned and again looked right at me.

"What would it be like if we could see what other people are thinking?" she asked, repeating the question I had asked.

"I'm going to let you folks in on a little secret," she told us while keeping her smiling eyes focused directly on me.

Peachey was one of those rare people who had a warm, familiar smile that, no matter how many people were in the room, she made it seem like she was talking only to you.

"I have been doing consulting work for Bill Dobelle's artificial vision labs in Palo Alto the last couple of years," she continued, "where we actually *are* beginning to get a few more images from the brain. Researchers at Dobelle are beginning to see a little more of what the homunculus is doing! Although they are not exactly sure what to make of some of the images yet, we know they are definitely related to some kind of 'thinking.'"

Peachey reached into her briefcase, which she had propped open on a chair beside her.

"I'm going to record the little lecture I'm about to give you today," she went on. "You'll be the first people outside of Dobelle to hear about this portion of my research. Feel free to ask any questions you like as I go along. But don't forget you're being recorded."

Placing her phone on the desk, she winked at us, saying, "The stuff in this recording should get me in the headlines of the *New York Times* someday."

Peachey held up her phone so we could all see it and hit the Record icon. She straightened her lapel, glanced at me again, and went into formal lecture mode to give it her professional sound and feel.

"Good morning again, everybody, you know I am Professor Edythe Peachey, lately with American Nonlinear Systems—I'll just refer to it as Nonlinear. I must make note of my thanks to the very forward-looking funding efforts of Patty Stonesafe at Nano-soft for seeing the potential of developing *Living Windows* software. Besides being at the Nano-soft Skunk Works in Redmond, as of this morning I now hold the Living Windows Nano-soft chair in computing sciences and psychology down at Stanford University."

"*Living* Windows?" interrupted a pretty young woman sitting two chairs on my left. "You have learned how to give Windows *life*?"

"All I can tell you today is that Living Windows is a whole new approach to software that is based on the homunculus you have just seen in the video clip," Peachey enticingly replied. "You'll have to attend my closing session tomorrow to get the details. But everything in this little lecture is interconnected. As I move on, see if you can figure out how we might be using the homunculus to give life to Windows.

"At Nonlinear I have been given a free hand to explore some pretty kinky ideas about the brain," Peachey continued without skipping a beat. "First, let me address Betty Jean's question about seeing what people are thinking. Some of our newer findings are coming from brain areas right next to where Pauline's homunculus migrated to the holographic, image-gathering electrode. These areas are where vision begins to connect up with what the rest of the brain already knows from past experience."

"Which brain areas are these?" asked a shy, Bill Gates–looking young man sitting just to the left in front of me. "Are they brain areas 18 and 19?"

"Ah, we have a brain scientist with us today," replied Peachey. "Yes, they would be areas 18 and 19. But the rest of you don't need to worry about brain area numbers."

"Yes," interjected the Bill Gates look-alike. "These secondary visual areas 18 and 19 constantly contribute updated meaning to incoming visual information."

"No matter what you call them, these are very important areas of the brain," Peachey went on, giving "Bill" a thumbs-up. "Areas 18 and 19 feed not only vision to the homunculus areas of the brain, but minute by minute, they feed thoughts, ideas, and past feelings to the visual world of the homunculus. It's where your homunculus and 'you' really come to life as your attention changes from moment by moment. Your ongoing train of thought has to be constantly knitted and reknitted to form your tapestry of conscious awareness. One of the early psychologists taught us about this 'stream of consciousness' over a hundred years ago. If the door to this room suddenly flew open with a loud bang, you'd all have to instantly switch attention, reload, and evaluate what was happening.

"Even when this young man asked if I was talking about areas 18 and 19," she continued, pointing to the young man, "you had to switch attention, reload, and evaluate. This is what I am talking about. Of course, other parts of the brain are working on this ongoing train of thought too, but 18 and 19 is where it all comes together thousands of times every day.

"Betty Jean," Peachey said, shifting her attention to me. "Remember the huge mouth, lips, and tongue of Pauline's homunculus?"

"How could we forget them?" a researcher in the back of the room yelled.

"When we move the holographic, image-gathering electrode to include areas 18 and 19, those huge mouth and lips of Pauline's homunculus move sometimes as if it were talking to itself," Peachey continued on without taking her eyes off me. "We had noticed that the homunculus's lips move sometimes when it is doing something, like dressing a mannequin, and sometimes when it is just sitting or standing motionless. Our lead lab technician, Jay Headley, happened to be dating a deaf student at the University of Washington in Seattle. They met in a martial arts class Headley was taking at night. Her name was Kimberly. Headley suggested that since Kimberly could read lips, we should get her to take a look at Pauline's homunculus in Fullview to attempt to read its lips. To make a long story short, she told us that indeed, Pauline's homunculus was talking to itself about what it was doing but also that when the lips moved when her homunculus was motionless, it was talking to itself about the experiment and other things Pauline was thinking about."

"So in a way, you *have* been able to see what Pauline is thinking," I interrupted.

"Yes, but only about half of what is being said can be deciphered by reading the lips and tongue," Peachey replied. "We are only at the beginning of finding out what Pauline is thinking by reading those lips jabbering away inside her brain. We now have lip-reading computer software that can read lips better than any human lip-reader. But we are keeping the details of this new development very quiet for now."

"But if you can get half of what is being said, isn't that enough to get the context of Pauline's thinking?" I asked.

"Yes, it is," Peachey replied, sounding a little like she wanted to move on. "With the lip-reading software, we are getting way more than half. There are strange lip movements that are like those of animal calls that intrude into her speech from time to time. Again, as I said, we need to keep this very quiet for now."

CHAPTER 5

The Walking Dead and Planet of the Apes

"Let me reveal to you some of the most startling things we've very recently discovered about the brain," Peachey hastily went on. "I am afraid that some of you will perhaps find these ideas a little bizarre, but please bear with me. I'll start with a small shocker for you and then work my way up. Our team at Nonlinear has discovered that there is a whole storehouse or archive of images in our brains that were put there over the course of at least the last two hundred million years of evolution! Personally, I can't think of any scientific findings that are more thrilling to contemplate. People, this means that the most basic archive of images that each of us carries around in our brain has been passed down from the dinosaurs! This dinosaur archive was established in our brains across literally hundreds of thousands of generations of living creatures! This 'reptilian' archive of images is the distilled experience of everything the dinosaurs felt and did—EVERYTHING! Think of it! Remnants of the sights and sounds and, you'll see in a moment, powerful urges starting with the dinosaur era are stored in ancient image archives that live in our brains. This finding alone is enough to intrigue the hell out of scientists, including paleontologists, for the next one hundred years!"

"Dr. Peachey, we have heard about the human homunculus images you have discovered in your lab, but how do you know our brains actually contain dinosaur images?" asked a geeky researcher sitting way in the back of the room. "I mean, sure, it makes sense that we would have homunculus images in our brains. Those images are the ones that actually feel and do everything for us in the *now*. They knit the tapestry of conscious awareness together, as you put it. But dinosaur images in our brains? Why would we need ancient dinosaurs?"

"Good questions," Peachey went on. "First, how do we know about these images? It's quite simple. We have seen them!"

"What!" exclaimed the geek-looking researcher in total disbelief. "You have *seen* them?"

"As I was just saying," Peachey continued with a smile that understood the disbelief, "through the kind cooperation of Dr. Dobelle, we have been following some kinky research ideas that have gone beyond areas 18 and 19 and have been using the holographic image-gathering electrodes to probe the entire brain for more images. The 'dinosaur' images we have discovered have migrated to image-gathering electrodes we placed in a very old part of the human brain called the R-complex. It's called the R-complex because it is made up of ancient *reptilian* brain tissue. We have known for decades that it is definitely dinosaur stuff because when these reptilian tissue assemblages are chemically stained, they turn the same colors as those of today's nearest living relative of the dinosaurs. That's the giant Komodo dragon of Indonesia."

"The Komodo dragon is a dinosaur?" I blurted out.

"Oh yes, very much so," Peachey replied.

"The Komodo dragon evolved at least one hundred million years ago and hasn't changed since," she continued. "The Komodo dragon was a contemporary of the great *Tyrannosaurus rex*, but the T. rex died out some sixty-five million years ago. Due to its isolation when the dinosaurs were mostly wiped out, the Komodo lived on. Using the chemical-staining techniques, it is fascinating to go into the *human* brain and see that these ancient reptilian tissue assemblages are still *alive* and functioning in a perfectly preserved state. The second question you asked was why we need these dinosaur images. Why are they still there? The answer is simple. The reptilian complex runs all of our basic instinctual functions, just as it did for the dinosaurs. Every creature has to have them in their brain to live, fight, and to procreate. You know, the Komodo dragon will eat almost anything, and it drives that same urge in humans. The dinosaur images our staff has gotten so far are a little shadowy and grainy. Unfortunately, this means that we haven't been able to do much-detailed study on them.

However, we are presently working to refine the placement and shape of our holographic image-gathering electrodes."

"You're refining the shape of the electrodes?" someone in the back questioned. "How does that help improve the images you are getting?"

"We are not yet sure how the shape improves the images," Peachey quickly replied. "We happen to notice that some of the image-gathering electrodes gave us better results. When the electrodes were examined under a microscope, we found that the ones that gave us the best images were ones that had rough micro burrs on them. We think the micro burrs gather more details of the holographic images, so we are maximizing the number of burrs.

"At any rate, the increased number of burrs on the electrodes is already showing signs of giving us a better view of the ancient reptilian world," Peachy continued. "Can you imagine? With further image clarifications, we might eventually be able to see how the dinosaurs' homunculi saw the world hundreds of millions of years ago. Think of it, 'living paleontology.' How much better to see the dinosaurs from the inside out rather than just the fossil remains of their outsides! We are at the beginning of seeing not just bones of dinosaurs but the actual mental images of the urges of dinosaurs that form the base of our minds. Dinosaurs are buried not only in the earth but in our own minds."

Good god, I thought, *it's bizarre to think that we have living dinosaur tissue in our brains. People run around all over the world, trying to find skeletal remains containing viable dinosaur DNA, and all the while dinosaurs are alive right in our own brains.* "Is this why we were so thrilled at the sight of bringing them back in movies like *Jurassic Park*?" I muttered to myself.

"I'll bet that's why we like dinosaurs so much," I blurted out. "Even as kids, we all want to go back to the land of the dinosaurs. Maybe we are fascinated by them because part of our brain somehow recognizes them. They are part of us, just as our homunculus is. Looking at dinosaurs stirs an ancient familiarity living within our brains. It brings that part of that ancient part of our brain back to life for a moment. Maybe we really can go back to the land of the dinosaurs! Maybe Pauline's homunculus has already met the dinosaurs living deep in her brain!"

Suddenly, I felt I had just made a very stupid and naïve-sounding little speech; I quickly put my hand over my mouth.

The room was quiet. The people sitting near me leaned forward to get a good look at me. You could hear a pin drop. Peachey looked at me, gently nodding her head, peering at me pensively as though there was a growing connection between us.

She returned to her lecture.

"Some of you are looking just a little dubious, or maybe you're confused, about all of this. Let me back up a little to the research that originally led us to go looking for the dinosaur images in the first place. How many of you have heard of Paul MacLean's pioneering work at the National Institute of Mental Health's Neuroscience Center? I thought so—none of you. MacLean was a brilliant neuroscientist. He found that the human brain actually consists of three different kinds of brains that developed over the last two hundred million years. Each of the three different brains originated over this great expanse of time to allow animals to do different kinds of things in an ever-changing world so they could survive. As you can guess, each of the newer brains developed out of the older ones, and they were all designed to work together. But under certain circumstances, because they originated as different animals, they can be at severe odds with one another."

"So what happens, Dr. Peachey, when they are at odds with one another?" asked the young man sitting next to me. "I mean, how do we know when the different brains in our head aren't getting along?"

"Let me tell you a little more about the three brains MacLean discovered before I answer that," replied Peachey. "The more you know about the three brains, the more it will make sense to you.

"As each new stage of the brain developed across 200 million years, the older brain had to be retained to continue to carry out its survival functions," Peachey continued. "This brief lecture doesn't give me enough time to go into a lot of detail today, but here are the barebones of MacLean's discoveries. MacLean found that deep inside the human brain is an ancient reptile's brain, which takes care of basic instincts—for example, for self-preservation, feeding, sexuality. The reptile part of the brain began with the dinosaurs. Next, situated on top of the reptile brain, he discovered there is a mammal brain, the brain that is dominant in caring for the young, social bonding, playfulness, joy, and pleasure in dogs, horses, lions, and of course, apes and humans. Finally, the third and most recent development is the uniquely human part of the brain. This newest part of the brain allows us to accomplish abstract reasoning and human language. This uniquely human part of the brain is somewhere between three million to a *mere* two hundred thousand years old."

Dinosaur brains, ancient horse brains, and ancient ape brains all buried in my *brain*, I thought. Strangely, this thought gave me a sense of belonging to nature that I felt would be with me forever.

"Now, we can talk about what happens if these brains are at odds with one another," Peachey continued. "Probably everyone here today has felt what it is like to have their three brains not quite working in sync," replied Peachey. "Think about the last time you got mad and yelled at someone only to have calmed down a few minutes later, wondering why you were even mad in the first place. This common experience is like there are three different people or 'voices' in your head. Here's where those three voices came from. First, something happened that threatened you, and in a flash, you became aggressive—that's your dinosaur brain reacting. Next, within a few milliseconds, you felt mad—that's your mammalian brain experiencing emotion. Finally, you slowly calmed down as the threatening situation went away, and you began to reflect and wonder why you were even mad in the first place—that's the newest part of your brain, the uniquely human part experiencing a disconnectedness from your dinosaur brain. These kinds of three-part brain-image flows go on in our brains every day of our lives."

"Does the dinosaur brain ever get seriously out of sync with the rest of the brain?" asked the young man sitting next to me. "I mean the dinosaurs sure got out of control in *Jurassic Park*!"

A few people laughed, but most of the room was silently waiting for Peachey's answer.

"The dinosaur brain actually does get badly, very badly, out of sync," replied Peachey. "Some psychiatrists believe many severe mental disorders are caused by the brain's ancient dinosaur images trying to take over and dominate the rest of the brain. The dinosaurs in us make us obsessive, compulsive, over-ritualistic, zombie-like, and sometimes, this obsessive, zombie-like behavior is a combination of primitive sexual and primitive feeding—all of these are dinosaur images in us."

"Uncontrollably sexual!" blurted out the young man.

"Yes, sometimes, it is uncontrollable sexual and feeding behavior combined," Peachey replied. "Some psychiatrists believe that madman Jeffrey Dahmer, who had violent sex with victims and then ate parts of them, was under the control of the dinosaur brain within him. No one else has been able to explain what could cause such bizarre, uncontrollable reptilian-type behavior in a human being!

"And curiously, back to Betty Jean's point about why it is we like dinosaurs," Peachey went on. "We are fascinated with *Jurassic Park*, and maybe more importantly, we are fascinated by movies about *The Walking Dead*. The people in *The Walking Dead* exhibit the zombie-like, out-of-control behavior of the dinosaur brain. We seem thrilled that they give expression to our hidden reptilian desires. And don't forget, maybe this is also why we love dogs and horses and why we are fascinated by our possible relationship to apes. And maybe too, it is the real reason why the *transition* between apes and humans is so credible in *Planet of the Apes* and *Rise of the Planet of the Apes*. We were, in fact, at that same stage of *transition* with all kinds of partly early humans and partly apes as they encountered each other in the struggle for existence for at least three to five million years. The mental images of that critical struggle are still living deep in our brains. The reason why we think of such bizarre things as *Planet of the Apes* in the first place and why we enjoy it so much is buried deep within us."

"Thank you for rescuing me from my stupid comments about why we like dinosaurs, Dr. Peachey," I said without waiting to be acknowledged.

"No, no, thank *you*, Betty Jean, for bringing up this whole issue of our obvious deep familiarity with our ancient animal past," replied Peachey. "This deep-running sense of connectedness is a whole new area modern science can explore, and we are beginning to do just that."

"Hear, hear," a few members of the session quietly murmured in agreement.

"Have any of the horse, dog, or ape images you mentioned migrated to your deep holographic image-gathering electrodes so you could get a look at them?" asked a professorial-looking type of very elderly woman off to my right. "I especially love horses, and that would be a 'kick,' if you know what I mean."

The room broke out in laugher at such an elderly woman's dry humor.

"Yes, we have seen the dog and the horse," Peachey replied. "But curiously, these images undergo strange transformations before our eyes, sometimes flickering back into dinosaur forms and sometimes into forms that look like a human homunculus—sometimes a female homunculus like Pauline's, sometimes a male homunculus. I don't want to put you off again, but if you can attend my closing session tomorrow, I will be able to tell you a little more about these very strange images."

"Dr. Peachey," interrupted the shy geek in the back of the room. "You told us about how Pauline's homunculus came to life where the electrode array was hooked to the brain. I can see how that image is *alive* because it feels and does everything for us every day, but what led you to think there would be other images still alive in the brain? What made you probe deeper into the brain, looking for 'living' dinosaur images in the first place?"

"Well, first, remember that the dinosaurs had homunculi just like we do," Peachey replied, smoothly continuing her lecture. "That's *homunculi* for any geeks we have with us. Only, of course, the dinosaur homunculus looked a little different from ours. Second, something I haven't brought into the discussion yet is the findings of the neurosurgeon Wilder Penfield back in the 1950s. Penfield observed that when certain parts of the brain were electrically stimulated, patients experienced memories of entire parts of their pasts. And these experiences seemed incredibly real to the patients, as if they were actually happening all over again. They could hear and see people they knew many years before, even in childhood."

"Are you saying the patients could actually *hear* the voices of people who were talking to them decades ago when they were children?" I asked.

"Yes, strange as it may seem," Peachey replied. "When a fifty-year-old patient's brain was stimulated in a certain place, the patient said, 'I hear my brother talking, and I can see him playing on the floor. He's asking me for a glass of water.'

"They can smell and feel the surroundings that go with those childhood experiences," Peachey continued. "Penfield concluded that every image we experience becomes a part of the brain forever—every experience we have ever had is permanently recorded in the brain!"

"Everything I have ever experience had been permanently recorded in my brain?" asked one of the graduate students who came up from Palo Alto with me.

"Yes, *everything*," replied Peachey with a determined certainty. "We reasoned from Penfield's findings that brain really is an extremely complex blob of tissue that evolved to permanently store and manipulate millions upon millions of images. After all, that's all we experience during our entire lives is—images, images, images, either real or imagined images—even a word is an 'image.' Karl Pribram, a leading neuroscientist, argued that all of these images are stored in very small detailed brain circuits as holograms. And so yes, there is room for these millions and millions of images in the brain. They really don't take up much space—it's been estimated that a chunk of your brain the size of a sugar cube could hold the entire contents of two hundred thousand novels. Images are packed so densely in the brain we call the world inside the brain nonspace. When we think, nothing really goes anywhere in the brain, and it takes no time to complete any distance of imagined travel. It really is nonspace."

Nonspace, I thought. *I wonder what Einstein would think about that.*

"Karl Pribram or no Karl Pribram, it is hard to believe that everything we have ever heard, seen, smelled, and felt is permanently recorded in mental images in our brains!" exclaimed a woman sitting behind me. "Do we really have complete memories of our childhood experiences?"

"Yes, I agree, it is hard to believe," replied Peachey. "The findings of the permanent record of memory shocked even Wilder Penfield. But there is much more to this story. Some of you may have heard of people who have *super autobiographical memory*. Technically, it's called hyperthymestic syndrome. These people can recall almost everything about their past—and I mean everything! One woman by the name of Jill Price wrote a book about her hyperthymesia. She said her memory was like having a video recorder on her shoulder of everything she had ever done. She could play these tapes and relive her past experiences in complete detail. Jill Price's memory plus the permanent memory records Wilder Penfield found in his patients show us that we all have complete, permanent records of memory. It is only when something in the brain isn't working right, like in hyperthymesia, that a rare handful of people can get access to them."

"You can get access to your permanent memories only if something *isn't* working right?" I asked, somewhat surprised. "It seems we would get access to complete memory if something is working right, not wrong."

"Yes, it does seem that way," replied Peachey. "But if you talk to Jill Price and others with super autobiographical memory, you find that such a powerful memory is not exactly a blessing. They tell you it really doesn't do them much good, and in fact, it often intrudes on their everyday thinking. At times, 'tapes' of complete, permanent memory will spontaneously play in Jill's head that interrupt her normal train of thought. Again, it is important to realize that we all have these complete, permanent memories but that the normal person simply can't access them. Our permanent memory archive is not meant to be accessed by the conscious mind. Because as Jill Price points out, a complete autobiographical memory of everything gets in the way of everyday thought, we think evolution selected a 'firewall' in the brain so that permanent memory could be accessed only by the unconscious mind, thus leaving normal, conscious thought processes free to solve immediate problems and to build new memories. That firewall is missing in people with super autobiographical memory.

"We're running short on time," Peachey said, looking at her watch. "Let me quickly summarize so we can move on. At Nonlinear, we simply combined the following two facts: first, all of the mental images we experience are permanently recorded in our brains, and second, we have ancient dinosaur brains inside our brains. Therefore, we must have permanent dinosaur images in our brains! We must have the dinosaur images to carry out the primitive self-preservation, aggression, and sexuality of our dinosaur past. As we go through our daily activities, the collections of images from all three of our brains are used like computer files to constantly reload appropriate past information as circumstances change—although most of the ancient dinosaur images normally operate at an unconscious level.

But they do not always operate on an unconscious level! I'll elaborate in a moment."

Suddenly, a very peculiar thought struck me.

"Dr. Peachey," I said as if at that moment we were alone in the room, "these mental images are *alive*, aren't they? I mean, the brain is *alive*. And the brain is made of millions of images. So the images themselves must be alive. There is an image of you in my brain right now, and so you are sort of *alive* in my brain, right?"

Peachey was silent for a long moment.

"Yes, Betty Jean, the images in our brains *are* alive," Peachey finally responded, again looking at me with that deep connection. "They are made by your brain, the brains of dinosaurs, and everything else with a brain. You are right, images you experience in your brain are alive exactly as the brain itself is alive. Mental images in the brain are the *living windows* of everything we feel, think, and do—*everything!*"

The idea that images and their thoughts and feelings were *alive* hit me like a complete Gestalt flip where suddenly you realize that you're not driving south as you had assumed, but north. Everything suddenly turns around. I had always figured that my thoughts about things were just *fictitious things*. But, but because thoughts are alive, they are far more than that. There's a whole holographic world of memories in my brain that is totally alive. It is actually the inner world that gives the outside world life! When I imagine an image of my boyfriend, even though he might not be present, his image in my mind is totally *alive*!

"I will give you a memorable example of just how *alive* mental images actually are," Peachey continued on as my own living images continued to run wild in my mind. "We are just beginning to see what the images that come from our dinosaur brains might look like. Surprisingly, the mental world stored in the dinosaur brain may turn out to be the most fascinating because it is the most intense. Just as the dinosaurs were huge and ferocious, their feelings of anger and pleasure were huge and ferocious. Ladies and gentlemen, let me take you on the longest short ride in history. Prepare yourself and remember, the terminology is all in the name of science! Images of purely sexual nature appear to begin in your reptile brain, where they first evolved for the most intense and savage procreation. Once started, these powerful surges of pleasure swiftly radiate upward to your other two brains, gathering up additional images of feelings and thoughts as they literally take over the brain. Every thought we have about sex has raced forward through millions of years in a quick repeat of evolution that takes us from the world of the dinosaurs to the present moment in only a matter of milliseconds. This super fast journey through millions of years is why sex is so—so what? So *fundamental*, so *surreal*, so *mindless* nothing quite captures it, does it? Well, maybe *mindless* does, but no matter how you put it, sex is really beyond mere words, isn't it? All you can really say is that you feel like you are definitely going someplace new but that you are also definitely along for the ride."

Peachey paused in her lecture and slowly walked over the window, leaning forward on the sill for a moment. She turned to us with a sneaky smile.

"Let me tell you about the Peachey Time Sled," she continued, raising an authoritative finger in the air. "The orgasm itself is definitely a super fast visit to the world of the dinosaurs just as it was recorded over and over again tens of millions of years ago. In an orgasm, sexual images and experiences of the dinosaurs are literally shot forward from the two-hundred-million-years-ago land deep in your brain to the present moment within a few milliseconds. The way I like to think about it is that an orgasm is actually a ride on a *time sled* that flies forward, if I may borrow a comparison from the physical world, at many times the speed of light! When the dinosaur within us smashes into the present moment, there's a burst of 'ancient light.' It's as if when the dinosaur suddenly sees the beauty of the light of day again, we experience an orgasm. Inside your mind, the dinosaur flashes open its eyes and lives again for a few brief moments. Time travel to the land of the dinosaurs that people have long dreamed about is not only possible but we all actually experience this type of time travel many times every day of our lives—but more about this in a moment. This is much more fun than riding around through *Jurassic Park* in a Jeep, huh? Next time you have an orgasm, think of me because you'll be riding the Peachey Time Sled."

Peachey stopped her phone recorder. She leaned back against the windowsill. The room was dead silent. Peachey stared off into space as if she had just heard her own speech for the first time.

After what seemed like a full minute had gone by, Peachey began giggling to herself and broke the silence. "Who wants a ride on my time sled?"

"With *you*?" one of the male attendees quickly quipped.

"Let me rephrase that," Peachey quickly shot back. "Who wants a date with a dinosaur tonight?"

The room broke out in wild laughter.

"Dr. Peachey, Dr. Peachey," the Bill Gates–looking young man interrupted as the laughter slowly subsided. "The Peachey Time Sled is very exciting, but is it possible that if ferocious dinosaur images were allowed to enter Pauline's holographic image-gathering electrode, they might get out of control? Could they end up doing something bizarre, like eating Pauline's homunculus, for god's sake?"

"That's a very interesting insight," Peachey responded, visibly taken aback. "We just can't answer such questions yet. I emphasize *yet*. Remember what I said a few minutes ago. We do think that the meanness and ferociousness of the dinosaur layers of our brains are responsible for bizarre, out-of-control thoughts and behavior like that of Jeffrey Dahmer."

"Okay, that's it, you guys." Peachey suddenly shifted gears, bringing the session to a conclusion. "Here's your assignment. Take a ride on the time sled. Go ahead, pair up. Just kidding! As you leave, pick up a complimentary copy of MacLean's chapter 2 on ancient cerebral functions then take a ride on the Peachey Time Sled. Ha ha. Next time we meet, we'll do two things. We'll finish my *lecture*, and then maybe I can get one of you to connect it all up for us."

Peachey glanced at her watch.

"It's almost three thirty," she said. "Let's get out of here."

As I walked across the Nano-soft campus toward my rental car, I was thinking, *Good god, I'm not going to be the one who "connects it all up." I knew that that time sled business would come up—how embarrassing! I haven't even been on the "time sled," as Peachey calls it.* I wondered if an orgasm really was a super fast two-hundred-million-year ride forward across time. Are there really secret dinosaur images still living in the *nonspace* inside my brain? Is it really the mental essence of the dinosaur that people experience when they have sex? Living dinosaurs, *my* living dinosaurs in Fullview—what a dream!

As it turned out, I didn't have to worry about telling anybody about my deepest secrets. That breakout session was the last time we saw Professor Peachey. The next time the session met, we had a new guest speaker. What happened to Peachey? We were told only that other commitments had called her away from the rest of the conference sessions. Our new speaker knew absolutely nothing about Peachey's work at Dobelle—or her time sled.

Later that night, as we disembarked from the plane back home in Palo Alto, I realized for the first time just how important the Nanosoft conference had been to me. It really did change my life. Knowing about the possibility that dinosaurs and ancient mammals still lived on in our brains changed everything for me!

CHAPTER 6

Living Windows

(September 18, 2014)

It's 9:00 a.m. in San Jose, September 18, 2014. I am about to keep a second interview appointment with American Nonlinear Systems. Dr. Peachey's lecture was playing in my head as it had done so often during the last couple of months. But I didn't know whether Peachey was still working with Nonlinear. I hadn't heard a word about her. This second interview was a follow-up for a position where I would participate in an experiment. The experiment could go on for up to six months. The first interview was really an interview for my brain. They had given me all kinds of paper-and-pencil psychological tests, even the Rorschach inkblot test. Then there were the *real* brain tests—the fMRI, a PET scan, and the EEG. The only initial qualifications for the job were that you had a so-called perfect brain and that you wouldn't mind having three small holes drilled in your skull. The pay for the perfect brain and for allowing it to be drilled was $25,000 for each of two subjects, one female and one male. This was pretty good six-months' money for me as a freshly minted PhD with lots of bills.

But I wasn't really in it just for the money. It was my understanding that Nonlinear was still doing major research on homunculus and dinosaur brain images like the ones that Dr. Peachey had talked about just a couple of months ago. And I must admit, I still wanted to know if the Peachey Time Sled idea was for real. I wanted to know what it would be like to actually see one of the dinosaurs living in my own brain. Maybe what I really mean is what it would be like to *be* a dinosaur *again,* if just for a few seconds! At any rate, if Nonlinear wasn't after those ancient dinosaur images Peachey had described, I wasn't about to sign my brain to the project.

As I contemplated the finer points of dinosaur life, I was surprised to see that it was an army lieutenant that would be leading me to my second interview.

"My name is Ann," the lieutenant said, introducing herself. "And you're Betty Jean. I have been expecting you."

Taking me gently by the elbow, Ann guided me quickly away. We entered an immense room that was open and lighted like the latest thing in auto showrooms. Ann's tall, thin body continued to firmly lead the way as we snaked by dozens of what appeared to be huge aquariums. Technicians were everywhere, jabbering away and intently staring into the aquarium tanks in front of them. I wondered why they would be studying fish at Nonlinear. The tanks were beautifully lighted, looking like the coolest aquariums that could possibly exist. As we passed down a wide main corridor among some really large tanks, I slowly began to realize that there wasn't any water or fish in them at all! One tank, about the size of Volkswagen Beetle, had two little airplanes flying around in it. The whole scene was in an unbelievable quality of 3-D! The planes looked so absolutely real I wanted to reach out to see if I could touch them. These were very strange little airplanes. For one thing, they didn't have any windows. And the airplanes' bodies looked like they had been covered with hundreds of glistening white sequins. In a slightly smaller tank next to the one with the airplanes, there were rapidly changing aerial views of the ground, again in spectacular 3-D. The aerial views looked like the ones I had seen when flying at cruising altitude on commercial airliners—they looked to be from about thirty thousand feet. I guessed that perhaps these views were being taken from the planes in the larger tank.

"What's going on with all planes in these incredible *tanks*?" I asked the lieutenant.

"Oh, those are integrated sentience 3-D programs," she said matter-of-factly. "The two different-sized image monitors, or *tanks* as you called them, are part of our Fullview project. The larger monitors show airplanes that are actually flying around at the Boeing test range in Montana. The smaller monitors are part of the same Fullview computer program, and it shows what the pilots and observers *inside* the planes are seeing. Everybody in the air and on the ground is aware of what everybody else is doing and seeing. That's why we call it integrated sentience—integrated sentience is aimed at achieving maximum shared awareness. We're trying to create a kind of a tech-assisted ESP, where everybody is aware of what everybody else is seeing and doing. Right now, as we look at the smaller tank, you and I are seeing *exactly* what one of the passengers is simultaneously seeing as they fly over Billings. In other words, Betty Jean, you and I are in an integrated sentience loop with that person."

"We are seeing *exactly* what one of the passengers is looking at?" I asked incredulously.

"Yes, *exactly* what they are looking at," Ann replied.

"The project is called Fullview for a reason," she went on, "The pilots and the rest of the people on board the plane can see everything in their airspace and everything on the ground. I mean everything! And *we* can see everything *they* see—it truly is integrated sentience. It would blow your mind if I could get you up in one of the Fullview planes!"

Ah, they are using a version of Fullview, I thought to myself. Peachey may still be with Nonlinear.

"We've got an extra ten minutes," Ann said as she motioned me to quickly follow her. "Maybe I can take you up."

We took an elevator down to a deserted basement that was barren except for a giant tube-like structure. The tube appeared to be made of a dark glass and was about the size of two skinny Greyhound buses put end to end.

"This is what's inside the planes that you saw flying over Montana in the big tank," she explained as we climbed a small flight of stairs and entered the tube. "It's a whole new kind of air travel experience. The goal, Betty Jean, is to put people into a completely invisible technology. Fullview is part of American Nonlinear's even larger invisible tech program. This baby is invisible technology that flies," she said as she rapped her knuckles against the beautiful, iridescent wall of the tube.

"What *is* this stuff?" I asked as I ran my hand delicately over the shimmering inner surface of the tube. "It looks miles deep. It's like looking out into space on a starry night."

"It's a Samsing Seamless," Ann replied. "Isn't it just beautiful?"

Ah, I thought to myself, *they have a version of the Samsing Seamless already.* Ann pushed a sequence of buttons on her iPhone. Suddenly, the tube came to life, and as open space rapidly spread around me, my attention was automatically drawn to the floor. I looked downward in shocked astonishment. There we were, suspended in midair miles high in the sky! I became faint and felt myself falling forward toward Ann.

"I'm sorry, I sorry!" I heard Ann saying as I began to come around. "I didn't mean to do that! Are you okay? I should have prepared you. I thought I was pulling in the Fullview integrated sentience feed from a plane sitting on the tarmac in Billings. Instead I accidentally patched us into one that's cruising at thirty-two thousand feet."

"I'm okay—I think," I replied as I regained my footing and my senses.

Ann quickly fiddled with her cell phone. Suddenly, we were once again standing on the shimmering floor of the Samsing Seamless.

"Now, we'll get the integrated sentience feed I intended to get," she said.

As before, the seamless, iridescent gray floor suddenly disappeared from beneath our feet. It suddenly appeared to me that I was standing about twenty feet above the basement floor with absolutely nothing to support me. I felt faint again.

"Here, you'd better take my arm for a moment," Ann said, tucking my hand under her arm. "I just pulled in an integrated sentience feed from another Fullview Samsing Seamless that is sitting on the tarmac in Billings. It's about to take off."

"My god!" I exclaimed. "We're standing midair over the tarmac?"

"Not really, Betty Jean," Ann replied. "We're still in the tube, and we're still in the basement at Nonlinear. Just keep in mind that what you are seeing is only an illusion created by integrated sentience and Samsing. But I admit that until you get used to it, it can be an extremely unnerving experience. The Samsing Seamless is engineered beyond the abilities of a person's senses to detect the illusion."

"It's sure way the hell beyond my senses," I replied. "I'm not sure I can take this."

"You'll be okay," Ann went on. "Just keep one eye on me for a few minutes instead of looking down, and it'll help you get past it. I know, it's kind of like being at the top of the Empire State Building, and suddenly the railing is gone and there's no cement walkway beneath your feet. I've been up in an integrated sentience plane many times, and it can be amazingly scary. You'll be up there at thirty-two thousand feet, and no matter where you look, all you will see is the wide open spaces. It's like a magic carpet, except there's no carpet. Sometimes I'd momentarily forget that there was an airplane around me, and it would scare the living hell out of me. The Samsing Seamless creates a perfect reality. It's so perfect you can even look at the Samsing Seamless with a pair of powerful binoculars and they work just like you were outside in the real world. This iridescent surface has the same visual depth as a perfect mirror that's been turned inside out."

I wasn't exactly sure what an "inside-out mirror" was, and I wondered if, as she was suggesting, binoculars really could be used while looking into a regular mirror. The slight puzzlement must have shown on my face.

"Yes," Ann said with a smile, "binoculars do work in a mirror. But of course, you have to use them on something *in* the mirror that's far enough away to get them to focus properly, just like you do when you're looking at anything with binoculars."

It appeared that the two of us were standing there, suspended in space with nothing around us. Our bodies began to gracefully move through the air as the plane rolled down the tarmac. It was good-god weird. There we were, suspended in midair, gliding down the runway faster and faster. The earth rapidly fell away, and the vista all around us grew exponentially. It was just Ann and I—no perceivable plane around us, no wind, and only the distant sound of our commercial plane taking off. As we rose toward the clouds in this invisible jet-propelled escalator, a subtle feeling of power and control entered my body and, strangely, I began to feel incredibly happy. Information was pouring into our brains in a new way. We had entered a new reality, and it was *fun*!

"Sit down if you wish, Betty Jean," Ann said as she appeared to lie suspended in the air with her chin stretched out across the backs of her hands.

I had to remind myself that she was actually lying on the floor of the Samsing Seamless, which now could not be distinguished from the ground several thousand feet below and rapidly receding further and further beneath us.

"This must be what it's like to be a bird," I said as I flapped my arms and thought of running forward through the "air."

"I want to run and flap my wings, Ann," I said.

"Go ahead and run, Betty Jean," Ann replied.

I ran forward some thirty feet with my arms flapping away. I was flying; I was really flying!

"It's our natural thing, isn't it?" Ann went on, sharing my excitement. "No one wants to be constantly reminded they're dependent upon technology, do they? People don't want to think they are *using* technology, they want to think they *are* the technology—that it's their own body, their own senses that is doing everything. I think deep down people want to feel that the technology they use releases the superhuman they really are. We want to live a dream, and it's impossible to tell the difference between a dream and the powers of truly invisible technology."

"That's pretty philosophical, Ann," I said as I sprawled out in the midair next to her and gave her a wink. "Invisible technology is kind of like the brain, isn't it? I mean, we never see our own brains either—it's invisible to us. We never see our neurons grinding away. We assume our experience is what we *are*. We don't think we're a *brain*. Our biological brain technology that we live in all of our lives is completely invisible to us. Maybe that's why invisible technology is the ultimate thing. It's like getting inside a bigger, more powerful brain. It's our familiar feeling of being a *mind* that somehow inhabits the machinery of the brain."

What a thrill! The Samsing Seamless had secretly slipped inside our skulls.

Ann and I glided up to what appeared to be the customary thirty-two thousand feet. We began paralleling the flight of another plane. It was covered with the glistening white sequins I had seen earlier in the larger Fullview image tank.

"It looks like it's about a mile away from us, Betty Jean," Ann said. "Shall we open it?"

"What do you mean, 'open it'?" I asked.

"It's part of Fullview," she replied. "We can open our other planes so that we can see what's going on inside, and we can see it in as much detail as we wish."

Ann punched a single number on her phone and, this time, put the phone to her ear.

"Bryce, this is Ann," she said. "I'm in Samsing Lab 2. Give me the open code for our sister plane flying toward Miles City, would you please? Thanks, Bryce. Not only can we see the world around us as if we are flying through the air without the airplane but we can also see all of the planes from a distance just as we could see them in the larger image tank upstairs. And we can 'open' any of the other planes so that we can see everything that's going on inside. We can open the planes from here in the basement too—it's Fullview's integrated sentience for anyone that's tapped in to the system. You do remember that we are still in the basement, don't you?"

"This Fullview stuff is amazing," I politely interrupted. "Good god, I've got a million questions! For one thing, how do you get the outside view of the plane for the big image tank?"

"Oh, that's pretty easy," she said with a smile. "American Nonlinear has the entire earth mapped within an electronic image grid. An exact replica of each plane that's in the air is electronically recreated—right down to its paint job. We can select any number of airspace quadrants from the grid and watch what the planes are doing in that airspace. The pilots in each of our planes can also download any portion of the image grid they wish. That way, they can see themselves and all the other planes around them at any time and from any vantage point—just like we saw them in the larger image tank."

"That all makes sense," I replied. "Boy, what a deal for air traffic controllers, huh?"

"That's right, Betty Jean," the lieutenant said. "When this part of Fullview is put into airports and planes, the job of air traffic control will actually be kind of fun for a change. Have you ever noticed that anything becomes fun if you just have enough information to work with? That's my theory of *fun*—information is fun, fun is information.

The whole point of culture is to make more information available, and that makes more fun for everyone!"

I must have been giving the lieutenant a strange look.

"No, I really mean it," Ann went on. "The Internet is fun because it's pure information. New movies are fun because of the new information that's poured all over you. Jokes are fun because the punch line hits you with a sudden burst of information. And in the same way, doesn't it make you feel good when you discover something new? Disneyland is fun because it excites your childhood with more information. *Anything* can be fun if its net effect is more information."

"Okay, but what about *bad* news?" I questioned. "That's information, but it's not *fun*."

"Oh, I think I can disagree that bad news is information." She winked back. "Bad news is actually an information blackout! When you get bad news, you don't know what to think, what to say, or what to do. You automatically search for meaning—why did this happen? Bad news momentarily slams the door on information. A lack of information hurts and makes us cry. It's just the opposite of having lots of information and lots of fun."

I had to admit, the lieutenant had a point. She looked at me, smiling and waiting for a response, but I had none.

I thought about the Peachey Time Sled. A ride on the time sled is for most people the ultimate pleasure and *fun*! When you're on the time sled, you travel two hundred million years of brain evolution in milliseconds. Traveling through two hundred million years of brain development in milliseconds is the densest experience of information available to us. Peachey says the result of this ancient rush of information is an orgasm. *Hmmm*, I thought, *is that why we humans are constantly seeking more and more information?* I chuckled to myself. Is the computer revolution really all about our sex drive? Is Nano-soft the sexiest company in the world?

"Okay, next question," I said, shaking myself from my thoughts of the time sled. "How do you get the view that we're seeing in the small tank? How do you get the aerial views that people on other planes are seeing?"

"Speaking of fun, this is where it really gets fun," Ann said with a smile. "First, remember that these deep, iridescent walls of the Samsing Seamless have the same visual depth as a mirror."

"The same visual depth as a mirror?" I said. "You brought that up earlier, and I wasn't sure I understood exactly what that meant."

"I was wondering if I was going to get away with that one again without an explanation," the lieutenant responded with a sheepish grin as she looked out at the clouds flying by. "Betty Jean, when you look into a mirror, it gives you the same 3-D depth that you get when you're looking at plain old reality—like when you're looking at me right now. If you're standing two feet from a mirror, looking at your face, your eyes and your brain think your face is two feet inside or *into* the mirror. Your eyes actually look two feet into the mirror to get a clear, focused image. I must admit, Betty Jean, that I didn't understand this idea at first either. Next time you're looking in a mirror, stand about two feet from the mirror and put a mark on the mirror where your nose is. Then if you look directly at the mark, you will find that your face is totally out of focus. You must look past the mark and *into* the mirror to see your face. A technician who calibrates the Samsing Seamless tubes showed me a neat trick about this. Remember a few minutes ago I told you that binoculars worked when you looked into a mirror? One day, he brought a mirror and a pair of binoculars to the lab. He took me outside and had me look into the mirror at a license plate on a car that was behind us by about a hundred feet. At that distance, I couldn't quite make out the numbers on the plate in the mirror. Then he had me look at the license plate in the mirror with the binoculars. To see the license plate in the mirror clearly, I had to adjust the binoculars as if they were focusing for a hundred feet. It's the same with the iridescent walls of the two-way retinas. You can look into them with a pair of binoculars. The iridescent walls create perfect depth—it's exactly like the depth you see in the real world."

"I didn't realize that we had to look *into* the depth of mirrors," I replied. "That's weird. I'll just have to try those two mirror experiments.

"May I ask one more question?" I timidly asked.

"Sure, we have a minute or two before I have to get you to your interview," she said as she glanced at her phone.

"I remember that you said that the scenes in the small tank were what a particular passenger was seeing as they flew over Billings," I said. "How are you able to see that?"

Ann pushed a button on her phone.

"Bryce, would you please go ahead and open our sister plane that is flying over Miles City for our guest?" she asked.

"Open 17," we heard him say without asking Ann why she wanted it open.

Suddenly, there they were, four passengers and two pilots suspended in the air at thirty thousand feet, appearing to be flying just beside us. It was unreal. It looked like something out of a dream—a very bizarre dream!

"God, Ann, Fullview really is a *full* view!" I said in amazement. "What we are seeing seems impossible. You actually *opened* the plane. It's like being a god or something. Before telling me how you know what a particular passenger is seeing, you've got to tell me how this *opening* is done. I'm completely dumbfounded!"

"It *is* pretty new stuff, Betty Jean," she replied. "It's only been two years since South Korean researchers discovered that high-definition pixels of television screens could be designed like a two-way retina. Like the retina of the eye, a television screen takes in electronic information from the cable or the dish. But an equal number of nano cameras can be sandwiched in between the normal pixels and trained at the viewer to see what that television viewer is doing, even exactly what that viewer is looking at. This is what the Fullviews are, two-way retinas. Total integrated sentience looks outward and inward at the same time. When you're a passenger in a Fullview, you're looking out through the Samsing Seamless, and the Samsing Seamless is looking *in* at you. You're right, Betty Jean, it does seem that only a godlike creature could look into the eyes of each person in a planeload of people flying at thirty thousand feet."

"Ann, is this why the outside of the airplanes in the big tank are covered with 'sequins'?" I asked.

"Yes, Betty Jean," she replied. "The *sequins*, as you call them, are nanocamera nodes that allow for extreme telescopic and microscopic viewing. The nodes are pooling points for optical and other electromagnetic data."

"That's weird!" I said. "Integrated sentience is watching what's going on inside the plane at all times, and the people inside see the world going by *all around them*. The people are watching the world, but the world is also watching them—*very* weird."

"Technology always teaches us new possibilities about how we can remake what we thought was reality, doesn't it?" the lieutenant said. "Now, Betty Jean, you might be able to guess how we know what a particular passenger is looking at."

"Bryce, give us a close-up of the eyes of one of the passengers, would you?" Ann spoke into her phone.

A view of the eyes of one of the passengers enlarged before us as if they were floating in space. I stood there dumbfounded, looking into the eyes as if they were part of a dream.

"Remember, Betty Jean," Ann said, "those eyes are being projected to the Samsing Seamless wall and only appear to be hanging in mid-air. Let's see what those eyes are looking at."

Ann touched the "eye-out" icon on her phone. Immediately, where the eyes had been, a view of downtown Billings began to zoom in. A house appeared then the front door of the house.

"As you can see, Betty Jean, right now, the eyes of that passenger in the back of the plane are looking at a house as they are flying over Miles City. Maybe that passenger is simply looking at his own house. Who knows?"

"It's amazing," I said. "How are we able to zoom in on something someone else in another airplane is looking at? Are we becoming gods?"

"No, we are not becoming gods," Ann said with a smile. "The nanocameras in Samsing's two-way retina locate the individual's retinas on the grid of the 'sequins' you see all over the outside of their plane. A microcomputer in that nanocamera node then keeps track of where those retinas go and computes exactly where the retinas are looking. The node then shows us exactly what they are looking at in the exact magnification at which the Samsing Seamless is operating at that point on its surface. Very simple, huh?"

"Yeah right, very simple!" I jokingly replied.

"What about bad weather?" I asked. "Do the pilots and passengers lose their view of—of *everything*? And do we lose our view of what they are looking at?"

"Actually, no, *they* don't, and no, *we* don't, Betty Jean," she replied. "The Samsing Seamless also receives a complete overlay of Lynx-3 synthetic aperture radar that covers everything. The Lynx-3 has an algorithm that converts radar imaging into high-definition color images."

"Lieutenant, I'm sorry," I interrupted, "but I will have to admit my ignorance. I've never heard of synthetic aperture radar. What is it?"

"No, no—I'm the one who should be sorry," she replied. "I shouldn't have thrown that one at you without explaining it. Synthetic aperture radar, SAR for short, is a radar imaging technique. It's a camera that uses radar instead of light to get pictures. The radar can penetrate any kind of weather. It has been providing NASA and the military with very high-resolution aerial images of just about everything for many years. The Lynx series of SAR developments has refined the resolution of radar images to almost unbelievable levels. For example, pilotless military aircraft are used for surveillance all over the world. Those aircraft carry the early versions of Lynx SAR. The newer Lynx-3 combines SAR with new computer algorithms, which were developed at an artificial vision laboratory. The algorithms turn the radar images into the high-definition images for Fullview. That way, Fullview gets its reality-quality visual images under any weather conditions, day or night. Lynx-3 uses those 'sequins' or pooling nodes that you see covering the outside of the planes."

"*Very* cool," I replied as we left the Samsung Seamless, went up the stairs, and began to walk on past the Fullview area. "So where is American Nonlinear going with all of this stuff? Is the general public ever going to get to ride in one of these Fullview planes?"

"Well, the big picture," Ann began, "is that this whole room of Fullviews is part of our development for *Living Windows* 2016 for Nano-soft. And Living Windows is part of our larger invisible technologies initiative. Nano-soft software designers think that commercial airplanes, trains, and cars will all use the Fullview system within another ten or twenty years. So Nano-soft is going to be expanding beyond anything anyone can imagine. People are really going to be excited when they find out that Living Windows 2016 will be the operating system for a Fullview version of personal computers. Of course, though, the PC operating system will no longer be called Windows by 2016. With Fullview, regular old Windows are definitely going to be sort of passé, aren't they? Samsing is building the Fullview iPhones and PCs. The Samsing iPhone will put people *inside* an integrated sentience two-way retina computer."

"Samsing, Samsing, Samsing!" I exclaimed. "My god, I didn't realize Samsing was completely remaking the world!"

"During your research project, you'll find out that they do a lot more than build TV sets and hot electronic gadgets," Ann replied. "Samsing, along with American Nonlinear, is the primary designer of Fullview. The Samsing two-way retina iPhones will be a reality-sustaining integrated sentience like the one in the airplane—only, of course, it will be smaller. When a person slips on *Fullview glasses*, it will become their total reality. And since Fullview glasses will immediately become invisible, the person will forget that they're *in* anything. Like you said a minute ago, Betty Jean, it will be like quietly slipping into a bigger, more powerful brain. And as you said, your own brain is invisible to you. It'll be just like that! You'll still be you, your personality won't change—you'll hardly know the difference. But when you're *in* the glasses, you'll feel like you've evolved into a new kind of human being, with a whole new mental world at your command. It will seem like your own personal memory includes the Internet and that your thinking is being conducted by the most powerful operating system in the world—it will be a total *living* integrated sentience!"

"Wow, I just had a small insight about Fullview glasses," I interjected. "Well, at least it's an insight to me. The two-way retina technology could be used to permanently wrap us in a whole new greatly enriched physical reality. That could make our brains more powerful. If a baby were born into this Samsing Seamless world, it could live its whole life in there and never know the difference. The child could grow up to be a physicist in Fullview, and even though this child studied all the laws of the universe from inside Fullview, the child would think Fullview was the real world. But because of the child's enhanced experience, this physicist would have studied an enriched version of the universe and could know things about the universe we cannot begin to imagine."

"Betty Jean, you are really a smart one," the lieutenant said with an impressed look. "I didn't know you knew about nonspace! One of our research chiefs talks about nonspace all the time, and she even thinks Fullview glasses will someday allow us all to explore the two hundred million years of images that are stored in our brains!"

That research chief Ann is talking about has to be Colonel Peachey, I thought. *She must still be here.*

The Lieutenant barked "Open" to an immense door at the end of the corridor that led from the Fullview area. The immense door slid open without making a sound.

"Here we are," she said as she motioned me into another wing of Nonlinear. "By the way, Betty Jean, that's a secure door that only recognizes certain voices. Don't expect it to open for you until it has been programmed for your voice. But once it has programmed itself for you, it will know it is you and that you want it to do something. It will not only open and close but you can say 'see through,' and its surface will turn into a high-definition screen. That way, if we are in a lockdown, we can see what's on the other side before opening it."

"Well, it's been fun meeting you Betty Jean," Ann said as she began to walk back through the doorway. "I'll let Dr. Peachey explain the rest to you."

Yes, yes, yesss, Peachey *is still here*! I excitedly thought to myself.

As I stood in the corridor, watching the heavy door silently close, I felt the air around me move as if someone had silently slipped up beside me from behind.

"Hi, Betty Jean, I am so glad that *you* are the female finalist for my project," I heard a woman say in a lighthearted tone.

"I'm sorry, but how did you know my name was Betty Jean?" I replied as I turned.

"How could I forget *you*?" the woman said, smiling and tightly grasping my forearm. "You were the brightest and prettiest of my *students* up in Redmond at the Nano-soft conference."

"Dr. Peachey," I blurted out, somewhat embarrassed, "it *is* you. I didn't recognize you in that uniform. You're in the army? I had thought you were a free soul of the 1960s era. What's happened to the long beautiful hair and your—"

"And my beautiful hippie clothes?" Peachey laughed. "Betty Jean, I have always been in the army. I was Colonel Peachey when you were in my session in Redmond. The uniform just didn't seem to fit the intellectual atmosphere of an advanced Nano-soft conference.

"By the way, my hair is still there," she continued, turning with a slight dance so I could see the braids neatly tucked on her head. "And by the way again, I *am* a free soul, army or not!"

Peachey quickly became more serious.

"I have been watching your progress, Betty Jean," she went on, motioning me to walk with her. "I was very pleased that you came out on top in the brain testing. Your brain profile was *absolutely* perfect, no signs of childhood trauma, no neuroses, no psychoses.

"I should mention," Peachey said, stopping suddenly in her tracks and gently squeezing my forearm, "we were wanting to give Pauline Quinn a chance at this job. You remember from my Nano-soft lecture up in Redmond, Pauline is the young blind girl with whom all of this wonderful Fullview brain stuff got started."

"Yes, of course I remember Pauline," I replied. "I especially remember that you and Professor Vandervoort had begun to understand some of what she was thinking by reading the lips of her inner body, her homunculus. God, that was so fascinating! And I remember you mentioning her dinosaurs. I will never forget Pauline."

"Pauline wanted to see more of her dinosaurs," Peachey went on. "But because Pauline's inner body had experienced so much confusion with the electrode array at the beginning, Professor Vandervoort decided the deeper dinosaur probes might be too dangerous for her, at least until we find out what really goes on with them and what they dig up in Fullview."

Smiling and raising an eyebrow, Peachey poked me and laughed, "We should get nothing but clean dinosaur images out of you, BJ, and no special monsters will be created in Fullview from your growing up in that very *weird* Bay Area. Now, tell me, Betty Jean, do you still want a date with a dinosaur?"

We both threw our heads back in laughter as we mutually and fondly remembered Peachey's Redmond talk about her time sled.

Dinosaur images, I thought to myself, *that's what I have been dreaming about for the last three years!* Since I now knew that I had both Peachey and the dinosaur images, it looked like I'd be handing my brain over to American Nonlinear Systems.

As we continued down the corridor and then across an open courtyard within the building, I commented to her about the image tanks I had seen, "The Fullview project I saw in the outer lab is fascinating. Ann told me it was part of the Living Windows project."

"They're doing some pretty amazing stuff with that airborne integrated sentience," Peachey replied as we entered a warehouse-sized room. "But so far, the *living* part of the Living Windows project really hasn't gone beyond the fundamentals of existing computer technology."

We left the courtyard and entered what appeared to be a gymnasium. In the center of the room, a huge image tank loomed like something out of science fiction. It appeared to be about ten feet tall and twenty feet in length. In spite of its immense size, it was a glistening, more graceful version of the tanks I had seen in the Fullview area. Around it, scattered here and there in small groups, was a hodgepodge of technicians in white gowns, army personnel, and other professionally dressed people.

"But the project you're in, Betty Jean," Peachey went on, "is over a whole new threshold. Your project is beyond the leading edge. Nobody on the planet is anywhere near us on this one. In many ways, we're not even sure where it's going to take us—excuse me, I mean where *you're* going to take us. You are going into Fullview's holographic nonspace, where the image-gathering electrodes will read from deep in your brain. Your contributions to the Fullview Living Windows 2016 project just might lead to a new understanding of the whole history and meaning of human reality."

"Living Windows 2016!" I yelped. "What about the dinosaur images you said you were going to get from me? *Living Windows 2016!*" I repeated a little louder. "How are you going to get Living Windows out of dinosaurs? You'd *drill* my brain for Nano-soft?"

CHAPTER 7

Flight to Eternity: Talking Boxes in Nonspace

(San Jose, 2014)

Peachey smiled and put on a hand lightly on my shoulder and the other gently over my mouth as if to soothe and reassure me. She brought her face up close to mine.

"Honey, you'll get your dinosaurs," she whispered, mimicking the classic soft, raspy voice of Marlon Brando in *The Godfather*. "And the holes we're going to drill in your head are so small you won't even know they're there. This Windows project isn't going to be the kind of windows you're thinking about. In this project, we are going completely beyond all the old conceptions. For one thing, there won't be any *windows* in Living Windows 2016."

"No windows in Living Windows 2016?" I said with astonishment.

"No windows," Peachey replied. "As you think about different things, BJ, do you see any windows in your own mind? Are there any windows to open as you go from thought to thought?"

"I've never really examined what happens when I go from thought to thought," I replied with a mischievous smile. "Let me try it and see what happens. If I wonder what my mother is doing, there she is. If I want to think about the pyramids, they appear upon being summoned. I think canals on Mars, and there I am, cruising over the surface of Mars. Actually, Dr. Peachey, I'm not sure how I get from one thought to another—it just happens. And you're right, I don't see any windows in my mind."

"You never will see any windows in your mind, and there shouldn't be any windows in Living Windows 2016 either," she replied. "But we're calling the project Living Windows 2016 for security reasons. We want our competitors to think we are working on some kind of new *windows*. If other companies and labs knew what we are really doing here, we'd have corporate spies all over the place. Boy, is the world ever in for a surprise! Let me take you right to the bottom line. With the help of your brain, Living Windows 2016 will *think*! It will think just as you and I do. And so no *windows* will be necessary."

Good god, this is just beyond amazing, I thought. *The first software to actually think!* But how could people use a computer if it doesn't have some sort of windows? It seems like it would have to have *something* to open—a door, a hallway, a memory, an eye, a *something*! But Peachey's right; it is really strange that when I think about different things, nothing has to be opened. How can that be? How does the mind get from one place to another? The mind is so weird, I thought. *It takes you everywhere without going anywhere*. I wondered if Bill Gately at Nano-soft was in on this. I also wondered if he knew how our thoughts get from one place to another. I was sure that he must know that the end of the old idea of windows was fast approaching.

As we walked on past a myriad of technicians and scientist types, Peachey continued to reassure me that Living Windows 2016 was going to give me what I was looking for.

"Betty Jean," Peachey suddenly said in a deep, thoughtful way, "the biggest problem with developing Living Windows is for us to figure out why current computers can't *think*, can't think at all. We want to know just exactly what's deep inside us that allows us to think and why computers can't think. You're going to help me figure that out."

"Well, it's my understanding that computers just regurgitate or mimic stuff that we humans have already thought up," I replied. "Computers can't do any totally original thinking like we can, right?"

"Right," Peachey replied.

"Betty Jean, how much do you remember about Paul MacLean's book from my little lecture in Redmond?" Peachey went on. "I recall that you were very obviously fascinated with his findings concerning the dinosaur brain that's inside our brain. But how much do you remember about the details of his discoveries?"

"I remember MacLean's findings very well!" I replied with almost embarrassing enthusiasm. "In his brain laboratory at the National Institute of Mental Health, MacLean was able to convincingly show that there is a dinosaur brain and a mammal brain inside the human brain. I have been thinking about his work from the day you introduced us to it in your lecture. As far as I am concerned, his discoveries could be the most important scientific discoveries of all time. They could change the way we think about everything. I even think MacLean should have won the Nobel Prize!"

"If people ever completely understand the far-reaching implications of MacLean's work, he probably *will* receive the Nobel Prize," Peachey went on. "It is possible that our work on Living Windows 2016 will help him get the recognition that he deserves. I knew Paul personally, and he worked on his brain research every day in his lab in Bethesda into his late seventies. If he wins the Nobel now, it will have to be awarded to him posthumously."

"But I still hope he wins it," I added. "I have been thinking about the reptile and mammal brains almost constantly for the last year. The different animal brains have even shown up in my dreams, and I look at animals in an entirely different way now."

Peachey stopped in her tracks and looked at me, smiling playfully.

"Let's talk about this a little," she said. "This is where you can begin to help me. What exactly do you mean when you say you dream about the different animal brains?"

"Oh, they're mostly dreams about animal urges and feelings," I replied. "I don't actually dream about animals themselves. It's more that I have dreams about what it must be like to actually *be* an animal—if you can understand what I mean. And I think the animal feelings are making me begin to see why some of my dreams don't make sense. Our psychology professors always told us that dreams often don't make sense because something Freudian is being hidden from us. But I don't think anything is being 'hidden.' I simply think my animal feelings are allowing me to see that many of my dreams might be coming from the animals' point of view. I mean, maybe dreams are just the way animals feel and see things."

"But how would you know that these are *animal* feelings that you have in the dreams?" Peachey interrupted.

"It's hard to put my finger on it," I replied. "But in the dreams, I am convinced I am living through the mind of an animal. It's a matter of . . . of simple insights that I have in the dreams. I guess insights are feelings, aren't they? And there are absolutely no words for the feelings. In these dreams, it's like I don't know words or even know *about* words. When I awake from these animal dreams, it's like I had visited another world. I have to struggle very hard to find words—if I can find words at all—that fit the feelings I had in the dream."

"Maybe you do visit another world in your dreams," Peachey said. "I am very interested in just where you were in these dreams. You are saying there are no *words* in it and no *words* could describe it. That is very interesting because animals have no words! All I can tell you now is that this wordless animal world might tie directly in with Living Windows 2016. But go on. What do you think these animal dreams mean, BJ?"

"I've thought about that a lot," I replied. "And here's the gist of it, Dr. Peachey. In dreams, I think we often see the tapestry of the wordless animal minds that live inside of us. When we dream, it's human content, all right, but it is human content seen through the minds of the animal brains MacLean described. These feelings and images in an animal's mind are all going on at once, nothing is held back. To us, their world would appear to be a strange blend of deeply connected memories and events. It's far less ordered and less logical than ours. Maybe that's the way their everyday *thoughts* are—if I can call them thoughts. I know that it is only the most recent developments of the human brain that give any logical order and sequence to the world. In our dreamworld, our animal brains are free to express their animal drives and their ancient animal realities. We can't *be animals* very often when we're awake—except maybe when we're on the Peachey Time Sled!"

"My time sled made quite an impression on you, didn't it?" Peachey said as she interrupted my little speech and reached out to give my arm a friendly squeeze. "Please continue on, I am sorry I interrupted you."

"When we dream, the animals are allowed to surface, and we get to be an animal again while the recently evolved human part of the human brain is asleep and unconscious," I went on, smiling back at Peachey.

"That's an interesting hypothesis, BJ," Peachey said, raising her eyebrows in approval. "You just may be right. Animals do need to dream just like we humans do. Perhaps, as you say, the ancient animals in our brains still dream right along with the human part of the brain, giving it their pleasures and intuitions to enjoy and think about. To me, this is more proof that the dinosaurs aren't extinct after all. They are still *alive* and have simply become a part of our larger brain. The dinosaurs have an afterlife by being included in the brains of higher beings! When we dream, the dinosaurs light up the darkness in our sleeping brains. Their world is mixed with our world, and we get shadowy glimpses of the connections they saw and felt when we enter our dreamworld."

"Thanks for saving my dream theory," I said. "I thought it was too off-the-wall to have any chance of being accurate at all. By the way, I really like that—the ancient animals lighting up our sleeping brains. I think they do. And maybe in some distant time, our own thoughts will reappear to light up the dreams of an even higher being. Could this be why we have such a strong sense of an afterlife? The dinosaurs, the apes, early humans, they all have an afterlife through the remnants of their brains that still live in us! Maybe it's a natural part of evolution for us to go on living in the brain of a higher creature or, maybe, a higher configuration of electronics."

"You really feel these animals welling up into your thoughts, don't you?" Peachey replied. "Always trust your dreams. I believe that there can be great truths hidden in every dream. Wrapped up in our dreams are millions of years of animal natural selection in relation to the physical laws that govern the earth. I think that's why some of the greatest scientific breakthroughs have come through dreams. Perhaps your dreams have given you such a breakthrough. I have always thought that anyone can be an Einstein in their dreams. The trick is to figure out when our dreams are telling us something important. I don't think anyone has accurately decoded our dreams yet—not even Freud. Since your animals are still *alive* in your brain, lighting up your dreams, maybe you will help us find the *living* software code we need for our Living Windows project. Maybe you will even become an accidental Einstein!

"I'm also very interested in the other thing you said," Peachey continued. "When we were talking about Paul MacLean a few moments ago, what exactly did you mean when you said you looked at animals differently now?"

"Yes, it's the whole animal thing," I said. "This is hard to explain, but it's connected to the dreams. After the MacLean part of your lecture in Redmond, I began to have a strange urge to visit a zoo—any zoo. I've always loved zoos, but this was different. So when I went up to Seattle about a month ago to do research at the University of Washington, I found myself at Woodland Park Zoo. As I walked around the zoo, listening to people talking, I had the weirdest thoughts. I began thinking that compared to the animals, human beings are somehow strange, *hollow* things. It seemed that it isn't the animals that are missing something—it is we humans that are missing something. I can remember looking at the people around me in a very detached way. It sounds silly, but it seemed somehow alien to me how they were dressed in clothes and jabbering away about things the animals could never understand. I thought about how we and the animals live in two completely different worlds. As I looked them in the eye and they looked back, I think I even went to that other world and *became* an animal for those few moments. Perhaps I can't really explain this, Dr. Peachey. Maybe I shouldn't even try. The feelings I had were so bizarre, and I'm not sure there are words for them."

"Of course, you *are* an animal," Peachey said without hesitation. "In your dreams and at the zoo, you were obviously in deep sync with your reptilian and mammalian brains. I think a lot of people have some of the feelings about animals that you had when they go to zoos. I also think it's why we have zoos in the first place. In animals we can visit ourselves as we were as evolutionary children. Betty Jean, I think you feel the animal presence more deeply than most people because you know about the reptilian and mammalian brains within you. When you are deep in that animal feeling, you can see humans much more objectively. You can see humans as they really are."

"Yes, you are right, I am an animal," I said with a devious smile and a wink.

"You said 'we and the animals live in two completely different worlds' a moment ago," Peachey said, urging me to elaborate. "Remember, Betty Jean, we are trying to tie your thoughts in with Living Windows."

"Seeing and hearing all those animals in one place gave me tremendous oceanic feelings of a sort of primitive longing everywhere I went," I replied. "It was like I had stepped out of my human bubble and into a *really* real world. I got the feeling that maybe the wordless tapestry of feelings and images of the animal mind is the way the world really is. I felt convinced that this animal world was where I belonged. It seemed that everything *human* was some sort of a big, made-up game—the human thing was a game just like a totally contrived video game that wasn't real. That couldn't be possible, could it, Dr. Peachey? I mean, I seem to be contradicting myself. How could the wordless animal world be the true reality? How could the human world not be real? Such a bizarre idea doesn't seem to make sense."

Peachey looked at me without saying a word. Her face had a look that perhaps I had discovered something important, something maybe she already knew. I could tell she didn't want to answer my question. She wanted to see how far I could go with it on my own.

"Please tell me more," she said, stroking her chin and smiling playfully like a friendly psychiatrist.

It took me a moment to figure out what I actually *was* thinking. Then the music boxes I had seen while I was studying in France for a summer popped into my head. Why was I remembering those fantastic little music boxes? All of a sudden, it dawned on me that the words of human talk were like the preprogrammed music that played on those boxes. The music boxes were amazing. They sounded as if they were creating real, living music—the music was executed perfectly. But somehow, because the music boxes lacked real, living musicians or orchestras, their music lacked *heart*, the heart that made music, music. Music boxes can only mechanically pass on what has already been written by real, living composers and played by real, living musicians. I was quite surprised at what came out of my mouth next, and I had no idea where the words were coming from. I also wondered *why* I didn't know where my own words were coming from.

"When we humans speak, we are like music boxes," I announced. "Our words can only mechanically pass on what our animal brains have already thought and created. We humans are talking boxes. No music box has ever written a tune that didn't come from composers and musicians, and no talking box has ever said anything that didn't originally come from the animal brains constantly bubbling beneath it. When we speak, we are mechanical *talking boxes*. That's it. That's what I felt up in Seattle at Woodland Park!"

As we were about to leave the gymnasium area, I caught some commotion in the huge image tank out of the corner of my eye. As I turned my head to look, I was shocked to see an extremely ferocious-looking animal about the size of a large chimpanzee inside the tank. I couldn't believe what I was seeing. The animal's head was enormous—five or ten times the size of its body—with a huge mouth bristling with equally huge teeth. Its hands were gargantuan. Then as my eyes scanned its body in more detail, I felt my mouth drop open. Jesus Christ! The damn thing appeared to be skinned down to raw muscle! I looked at Peachey in dismay.

"Talking boxes, huh?" Peachey replied while pretending not to notice the startled look on my face and quickly guiding me away from the large image tank. "In the same way that the pins on the turning drums striking the note combs in, music boxes can only play emotionless notes as they are already written, the words of talking boxes can only pass on what the animal brains have already felt and created—this is great! Just keep going, Betty Jean, I think you just might be getting close to tying in with Living Windows 2016."

Tying in with Living Windows 2016? How could I be getting close to tying in with Living Windows 2016 with talking boxes? Does Peachey somehow know where I'm headed with these talking boxes before I do? The music boxes started intermingling with the rush of images flowing through my mind.

"How could my crazy idea about talking boxes be tying in with Living Windows?" I asked.

"You'll see, Betty Jean, just keep going," Peachey replied with a smile. "I must see if you can get there all by yourself. So just keep going."

I wanted Peachey to tell me what was going on with the *chimpanzee* a moment ago in the image tank, but she kept moving on as if it had meant nothing.

"Your insightful idea that humans *without* their animal brains are only talking boxes is brilliant!" Peachey continued. "It makes me think that in some mysterious way, you already know about the latest things we have discovered about the evolution of words, the evolution of language. These new findings have not even been published yet."

"What *have* you found?" I replied excitedly.

"We have always thought that human language evolved directly out of animal cunning," Peachey said. "And now, using fMRI brain-imaging on animals in our lab, we have discovered that all words and language are just vocal *tags* or labels for images that animals had already been using to fake out other animals for millions of years—in other words, to be cunning. Words themselves mean absolutely *nothing*!"

"Words mean absolutely nothing?" I asked, somewhat shocked.

"Yes, Betty Jean, words all by themselves are meaningless," Peachey replied. "Have you ever noticed that a foreign language sounds like gibberish at first? That's because it really is gibberish unless you know exactly how it's tied to the archives of animal images in your brain. Without the archives of animal images to give a language life, it is nothing but a labeling code, a bar code. I think that's part of what you were feeling at Woodland Park when you said human beings seemed somehow 'hollow.' Language is only a type of a bar-code system for the two hundred million years of images stored in our animal brains. But, Betty Jean, the important thing for us to realize is that bar-code systems don't think. They cannot do the most important thing in life—they are not creative. Being a bar-code system of sorts, language on its own cannot be creative and, therefore, neither can the talking box part of the human brain."

"Dr. Peachey, if words are like the bars or lines in a bar code and mean absolutely nothing, why did language evolve in humans?" I asked. "I mean, how did language even come about?"

"Let me give you some background," Peachey replied. "A moment ago, I saw that you were a little shocked when you noticed the homunculus of Gizmo, our lab chimp."

"Yeah, that was frighteningly strange!" I replied. "I was wondering if you were ever going to tell me what was going on there. So that was a chimpanzee's homunculus. It looked so ferocious!"

"Yes, they *can* be ferocious," replied Peachey. "But Gizmo is a very docile chimp. You remember Pauline's homunculus. Just like the human homunculus is the actual, strange form of the human inner body, Gizmo's homunculus is what the chimpanzee's inner body actually looks like. A lot of a chimp's brain is devoted to the hands, jaw, and mouth."

"A lot is devoted to their nipples too, I'll bet!" I blurted out. "Well, I mean they do spend a lot of time nursing their young, don't they?"

"Yes, yes, they do," laughed Peachey as she caught the humor. "That's what the mammal brain is all about—mammary glands! And I suppose, as in humans, there are also a lot of interesting brain connections with their nipples.

"Anyway, chimps have prelanguage areas in their brains," Peachey went on. "With our remote brain-imaging sensors, we noticed that when chimpanzees are trying to deceive other chimps about where they are hiding their food, their prelanguage areas become very active. Apparently, when chimps are trying to deceive, their prelanguage 'talk' simply consists of switching images around so they can try out different scenarios in their heads that might fool other animals. We have found that Gizmo's homunculus mumbles to itself when he is moving stuff around, his food trays and toys. Talking to ourselves is the same sort of thing. Talking to ourselves is nothing but switching images around in our heads, and talking to others is nothing but trying to switch images around in their heads."

"So this is a clue as to why language evolved in humans?" I asked. "You're saying that words could be used as vocal tags for images, and that way, early humans could quickly fabricate and share deceptions, right?"

"Yes, we think so," replied Peachey. "One of our own social psychologists working on the 2016 project happened to stumble on to something that we think is quite related. He had been analyzing thousands of samples of cultural communication activities from all around the world—you know, the Internet, newspapers, television, books, and so forth. This communication is the cunning *talk* of the 'talking boxes,' as you put it, which constantly goes on around the world. He found that 73 percent of all of this human *talk* is based upon deception. Most advertising, most politics, most of the stock market, most movies, most novels, most video games, most religions, and on and on are based on people using words to mislead people through deception. According to him, every conversation you have is 73 percent deception."

Suddenly it hit me.

"The epic myths, the gods, the ancient religions were mostly deceptions, and—and at the same time, they all were the most fundamental, the most complex examples of the use of words," I started rattling on. "Homer was the finest 'talking box' of his time. Playwrights, novelists, video game creators, witch doctors, politicians, lawyers, prophets—they have the finest talking boxes, and their words are most often about made-up worlds, worlds that don't really exist! But somehow, it's what we like best. I wonder why we like made-up worlds so much."

Peachey looked at me quietly for a long moment.

"There's more, isn't there?" she said as if she knew what was coming.

It seemed Peachey again somehow already had had all the thoughts I was having. Now she's watching me take that same journey. But how could she know this would happen? As I talked, I felt a little like Mozart trying to stay ahead of a piano concerto that was flowing forth from his mind. All I had to do was speak, and out it came. It was as if my own talking box was fast-playing the entire evolution of my animal world surging upward from below. I wondered how far it could go back in its animal history. Is this what Peachey expected to happen? Is this how she thinks I will tie in with Living Windows, if I just keep talking about "talking boxes"?

"Yes, there's more, Dr. Peachey," I replied. "A lot more. But is all of this making any kind of sense to you?"

"You're not going to believe this, Betty Jean," Peachey replied with a smile. "It's not only making sense, but I totally agree with everything you've said about the talking box. You're on a fantastic intuitive roll, honey, and you're going to love what I am about to say. There actually *is* a talking box, as you put it, in our brains. And you're right, it is a bunch of deceptive circuitry in our brain that evolved because of its tremendous advantage over the other animals. You asked why we like made-up worlds so much. That tremendous advantage is why—it gives an immediate feeling of power and safety! The talking box evolved to quickly create fake worlds out of strings of images that the other animals couldn't deal with. All of the talking box's stories are stories within stories within stories, all layered into what we call reality. But it's a fake reality that only we understand. And the bigger and more far-fetched the story is, the better. Think of all of the crazy ideas the ancient Egyptians had, but, crazy or not, their fake, talking box world mobilized them to build the most beautiful and powerful empire that had ever existed. The talking box mobilizes power, because, right or wrong, it puts everybody on the same wavelength."

"There's really a talking box in our brains?" I asked.

"Yes, there really is," Peachey replied as if poised to give another university lecture. "Remember Gizmo, the lab chimp? In the course of evolution, the prelanguage area in the brains of chimps became an actual talking box in the human brain. For now, just let me tell you that the *box* consists of several parts of the brain, especially brain areas 22 and 44. You can think of the edges of the box as being *drawn* together by a huge bundle of nerves called the arcuate fasciculus. It *is* sort of a talking music box in our brains—not too surprisingly, it's a box about the size of a small music box that would fit in the palm of your hand."

"Arcuate fasciculus?" I asked.

"Just trust me for now," Peachey replied. "You'll hear more details about this amazing little talking box later. Just out of curiosity, let me ask what could be sort of a personal question. A moment ago, I made the comment that for the talking boxes, it's the bigger the story, the better. What do you think about the Bible? Do you think it too is a *fake*, made up by the talking boxes?"

I still couldn't figure out where all my words were coming from. It was as if some autopilot deep within me continued to move my mouth. Maybe it was my homunculus moving its huge mouth. A realization was creeping into my mind that this whole blab fest of mine was triggered by the animal experience I had at Woodland Park. Somehow, I had come face-to-face with my own silly talking box while I was at that zoo, and I had seen through its entire charade.

"Is the Bible a fake?" I said, repeating Peachey's question. "What if I could prove that it is? That could be pretty scary, couldn't it?"

"Just let yourself go and tell me what comes to your mind," Peachey replied, urging me on. "Don't worry about the scary consequences."

I could sense the talking box in my own brain attempting to disguise my reply to Professor Peachey's question. Something was making it shy away from answering the Bible question head-on. I felt afraid to speak. Was my talking box afraid? Was the talking box afraid of what the animal world inside of me might reveal?

"God, I can't believe it," I said as an idea came into full view in my mind. "I have just had an intuition that means the animal world I experienced at Woodland Park really is the *real* world!"

I paused, with my mind stuck in the moment. I was speechless. Then I noticed Peachey was looking at me strangely. Her expression had changed to a distant smile, as if she was waiting for me to enter a world she already knew about. She seemed to know what was happening to me.

"Well, go on, go on!" she said, breaking the brief silence. "Tell me what you have discovered."

"The talking boxes think they are somehow ageless. They think they have always been *conscious*, and they think they always will be," I announced.

Curiously, I began to feel deeply sad.

"Ah, you're going to be such a good subject for our studies in Fullview," Peachey murmured. "You are showing good differentiation among the urges and feelings welling up in three brains within your own brain."

"Wait a minute, Dr. Peachey," I said. "If I am right that the animal world is the really real world, that means we truly are ageless. We actually have always been conscious! The talking box in me must be right about being ageless—my talking box is just passing on what the animal parts of me already know—they must be right about their living forever!"

"BJ!" Peachey said with astonishment. "Your mind, I mean your animal minds, are really cooking. And this is perfect timing to let me give you a brief rundown on how we believe we can get Living Windows 2016 to actually think. And if we can do that, maybe your poor talking boxes will be able to live forever—they actually are ageless! Later, either I or Professor Vandervoort will tell you about the All Time. You will love the All Time. It will help you prove that the talking boxes have always been conscious and will live forever.

And at the same time, you will find that what I am going to tell you supports your idea that the talking boxes know absolutely nothing!"

"Really?" I replied. "The most advanced part of the human brain is right that it will live forever but at the same time knows nothing? That sounds like the paradox of all paradoxes!"

"That's right!" Peachey said. "This is all going to sound bizarre at first, but you will see that it is true. Betty Jean, everybody knows that computer people have been trying for decades to build computers that can think, but even though computers can now do trillions of calculations per second, none of them has ever had one thought—not even a little one."

"Aren't today's supercomputers producing even *little* thoughts?" I joked. "After all, supercomputers have beaten some of the best chess masters."

Peachey raised one eyebrow and gave me a serious look.

"The idiotic way supercomputers have been used to beat chess masters has absolutely nothing to do with the computer doing any thinking," Peachey replied with firmness. "Supercomputers are used only as tools or *vehicles* by which programmers themselves are actually playing against the masters. And programmers are talking boxes who are creating another fake world, a world of computers that can play chess. Believing that computers are able to beat chess masters because they can think is like believing that cars can think because they know when to *go*, when to *stop*, and when to *turn*. If visiting Martians saw the complex traffic patterns of a typical city and didn't realize that there were people inside each car, they would have to conclude that cars were capable of an advanced level of thought! But of course, cars don't have even the tiniest thoughts. The people inside the cars do all the thinking. It's the same for *all* present-day supercomputers. Talking boxes make them do everything they do."

"I think deep down inside I have always known that computers don't actually think, Dr. Peachey," I said. "But I thought that computers were getting closer and closer to being able to think on their own. Now you're saying that they don't think at all?"

"That's precisely what I'm saying," Peachey replied. "And we don't believe that computers that are built like today's computers ever will think on their own—*ever*! The problem is that computer scientists have been trying to build thinking computers completely in the wrong way—basically, they have been going at it *backwards*."

"Backwards?" I said. "How could they be going at it backwards?"

"Everyone has thought that since computers can solve difficult problems in mathematics and, as you would put it, use languages as if they were like talking boxes, computers were getting close to being able to *think*," Peachey continued. "But computers are only mimics. We will never get computers to think just by making them better mimics. And that's what I mean when I say everyone has been going at it backwards. Thinking doesn't begin with the top of the brain, where the talking box is. It begins at the bottom with the ancient animal parts of the brain—the animal parts of the brain are the cranks that make the music boxes play and the talking boxes talk!"

"The animal parts of the brain are the 'cranks' that make the music boxes play," I repeated Peachey's words to myself. Suddenly, I found myself thinking about Peachey's time sled. The idea of shooting back tens of millions of years to the minds of the dinosaurs hit me in a new way. I could see a whole new reality waiting for us right inside the ancient part of our own brains. Peachey was studying my eyes as if she knew I was thinking about the time sled.

"Dr. Peachey," I said hesitantly, feeling as if she actually was looking into my thoughts, "were you serious about that 'Peachey Time Sled' stuff you talked about at the Nano-soft conference in Redmond?"

"Dead serious!" she said, looking me straight in the eye.

There was something about Peachey's tone that made me feel that the time sled might have something to do with the Living Windows project and Fullview. She continued to look at me, apparently deep in thought, without saying anything for a moment. I felt she was trying to decide whether or not to tell me more about the time sled.

"Well, at any rate," she said, trying to get us back on track, "I didn't make a point of it in my little Nano-soft lecture, but we think Maclean was able to show that *all* human thought has to begin in the dinosaur brain. Our dinosaur brain is what drives us to procreate. In other words, the dinosaur brain is what it means to *be*, to exist."

"How on earth can his research lead you to prove something like that?" I asked.

"As is often the case in science, the concept turned out to be very simple," Peachey replied. "However, to make that simple discovery, we had to perform the most delicate internal brain recordings ever attempted. MacLean had been mapping impulse patterns throughout the brain. The work consumed him for nearly twenty-five years, so we carefully retraced the nerve connections that led from the ancient dinosaur part of the brain, through the mammal brain, and on into the newer human part of the brain. As MacLean theorized, we found that the only time the human part of the brain could become involved in thought was when it was responding to impulses that started in the ancient dinosaur parts of the brain. It was *always* necessary for the nerve impulses to begin in the dinosaur brain."

"I remember in the Nano-soft lecture you told us that our brains are full of ancient images that may be as much as two hundred million years old," I replied. "But you didn't tell us that every human thought started in the dinosaur brain."

"We had our suspicions about it back then, but we weren't sure," Peachey replied. "We knew the images were in there, we knew how the dinosaur images raced forward on the Peachey Time Sled, and we knew they colored our human thoughts. We just didn't realize that they provided the framework for the origin of all human thinking. Every human thought is a dinosaur *thought* that gets *dressed up* as its impulses race through the evolutionary history of the nonspace toward the uniquely human part of the brain, where it is then expressed in human terms. The whole hundreds-of-millions-of-years history of the brain is retraced in a matter of milliseconds with every thought. And without traveling through all three stages of the history of the brain, there can be no thoughts—no normal thoughts anyway."

"So any thought that didn't travel through the *entire* history of the brain would be an *abnormal* thought?" I asked.

"Do you remember what I said about Jeffrey Dahmer at the conference in Redmond?" Peachey said with a serious look.

"Oh yes," I replied. "You told us that his violent sex with his victims and then eating them came right out of the reptilian brain. You said that's what happens when reptilian urges aren't modified by the later, more humane mammalian layers of the brain."

"Right," Peachey replied. "So getting back to computers, we've reversed the whole traditional, backwards approach. I have designed this entire lab with the idea that *all* thinking begins in the archives of images in the animal minds that MacLean's research revealed. The animal mind provides the basic driving *molecules* of every thought. People are never aware of the molecules in water when they take a drink, but without them there can be no water. It's the same with our thoughts. We're never aware of the ancient animal images within us that are driving and forming our thoughts. Computer designers have not been aware of this fact either. And if computers are ever going to think, they must have those image archives of animal urges and feelings built into them. To make a long story short, my dear, the *only* way we can get a computer to think is to give it the archives of living images of all three of the brains that are inside the human brain. We're going to get the living archives of images out of your *living* brain—we're going to get the *instructions* for building Living Windows out of all three layers of your brain."

"I knew it!" I trumpeted. "Paul MacLean really does deserve the Nobel Prize!"

Okay, I thought, *I'm beginning to like this Living Windows 2016 project. Just think, the world's first thinking computer is going to be some sort of copy of my brain. Ha ha!* As I was patting myself on the back, I noticed a small group of people approaching us.

"But how will it actually *think*?" I asked Peachey.

Waving to the group of approaching technicians with a fire-it-up motion, Peachey answered my question with a question of her own.

"Well, as I asked you before, how do *you* actually think?" Peachey asked. "In this business of designing computer software that thinks, I am always asking people to look inward to study what goes on when they think. Examine your own thoughts again and tell me what you see. Close your eyes and think about anything you wish. What do you see, what do you hear and feel?"

I closed my eyes and began to think of all sorts of things. My thoughts took me back to the Nano-soft conference where I had said, "Wouldn't it be great to see what people are thinking?" Other thoughts were of my mother and of our old house in Palo Alto. At first, I thought I was seeing faces and familiar rooms and so forth. But what was I really seeing? I examined my thoughts more carefully. The faces weren't really *faces*, like the faces you see on the street. They weren't faces like the ones I see in my dreams either. They were kind of jumbled lines—a line here and there, just sort of capturing an image of a familiar *face*. But as I looked into my mind more intently, I could see that these images had an order of their own. They have a deep order! "Deep order"—I didn't know where that phrase came from, but it fit what I saw in my mind.

Hmmm, I thought to myself, *I'd sure like to get one of these thoughts on canvas. I'd be famous! Wait, I know what deep order is. It's a mixture of all the layers of our two hundred million years of permanent memory. The deep order of faces is made up of all the animal faces that ever were and, at the same time, a particular face that I currently know. Peachey is right, my thoughts are all assembled and reassembled from my deep past, all the way back to its very beginning.*

Peachey interrupted my thought.

"Well, what do you see inside your mind?" she said.

"I am sort of—but only sort of—getting sketch-like images of scenes and sounds," I replied.

"What do you mean, 'sort of'?" Peachey asked.

"Well," I replied, "they're not really mental images at all, but I really don't know how else to describe them. They are somehow *disembodied* images, faces that aren't really faces, things that aren't really things—if that makes any sense. But at the same time, I know exactly what the images are. I think my thoughts are traversing here and there through the two hundred million years of layers of my permanent memory. It is a strange language of images. These memory layers are kind of like those Picasso captured in his *Guernica*. It's that kind of deep order that tells the true story of everything."

"Very good," Peachey said. "I too know of Picasso's *Guernica*, Betty Jean. It's an abstract mixture of a bunch of human and animal images put into disarray by the ravages of war. What you see in your mind, Betty Jean, is a similar language of images. You didn't describe the actual content of what you were seeing, you described something about the structure of your mind. The content is not so important here, it's the structure that's common to all thoughts that we're after. You've just observed some fragments of the structure of your own mostly unconscious memory. Your unconscious memory is made up of streams of what you have called a 'strange language of images.' This almost secret language is made up of small chunks of image *patterns* of sights, sounds, feelings, and so forth. These basic, ancient patterns underlay all of our languages, mathematics, and art. These abstract patterns can be combined in an infinite number of ways and allow us to learn any language, any mathematics, or any art—in fact, ancient cave people *read* cave paintings through these patterns. They are in everyone's mind to see if they look hard enough, BJ—just as you did. The brain's own ancient image patterns are far more beautiful than all of the flowers, sunsets, and intergalactic nebulae you'll ever see. They are the purest form of beauty. If you really want to experience ultimate beauty, you must look deeply into your own mind. When Albert Einstein described the deep order in his own thinking, he called it a series of such 'memory pictures.' You're an Einstein, Betty Jean!"

Aha! I thought. *What I saw as lines Peachey calls chunks of image patterns. These image patterns are the mind's basic instructions, and it uses them to think!*

"But," Peachey went on, "your personal unconscious memory is just the tip of the iceberg of thought. There's, of course, stuff even deeper and beyond Picasso. The dinosaurs had memories too, but their brains were very primitive. The image sketches of the dinosaurs' memories are coded deep within the thoughts of your own unconscious memory. The dinosaur memory images of sex, feeding, territoriality, defense, and attack, for example, provide the original power that drives your thoughts into existence. Beneath your every thought about everything—from all your questions, to your mother, to your childhood home—are the dinosaur images. To paraphrase Freud's real meaning of sex, whenever you produce a thought, you're having sex, you're having a remnant of dinosaur sex. Thinking is high-level sex, high-level aggression, and high-level ritual, high-level dinosaur! *Thinking*, Betty Jean, begins with an ancient code of images, and these images drive all the images of the new human brain, like the ones in your conscious memory, into existence. Thinking *is* this flow of *images*, and we must catalog and decode these animal image flows if we ever hope to build Living Windows software that thinks and *feels*. A computer will never be able to *think* on its own without animal images. I repeat, never! We already have the bar code part of the talking box in the computer. If we can somehow put the animal into the computer with that talking box, we might have figured out a way to give the talking box its ancient dream—this is the dream of going to the world of the All Time. There is talk that within a few decades, we humans could achieve a sort of immortality."

"The All Time?" I queried. "What exactly *is* the All Time?"

"Ah yes, the All Time," Peachey replied reflectively. "It's a beautiful concept developed by Professor Vandervoort. He has come to the conclusion that the human brain comes loaded with the last two hundred million years of images laid out in the layers of the permanent record of human memory. He calls this the All Time because he believes this two hundred million years of images is the only *time* we can really ever know."

"So the All Time stored in our brains is the only real time that exists anywhere in the universe?" I asked.

"Professor Vandervoort says that our regular sense of time is an illusion created by the 200 million years of experience stored in the three layers of our brains," replied Peachey. "The experience of time requires all three layers. So far, even our computers only have one of these layers, the last layer that we talking humans have. That's not enough to experience time. Professor Vandervoort wants to put all three layers of the brain into Living Windows. It will then have its first feelings of time passing by, its first feelings of the All Time."

"So it is because of the existence of the All Time that we know about time at all, isn't it?" I suggested.

Peachey paused, momentarily overcome by the meaning she had just experienced in her reflective thoughts of the All Time. A tear welled up in her eye.

"Again, we will have a computer that can think like a human," continued Peachey. "Even more, we will have a computer that can take us on journeys through the All Time, and in the All Time, Vandervoort thinks we can experience the dream of immortality."

"Immortality, is that possible?" I questioned.

"Hopefully, while you are in Fullview, you'll be the first to find out," replied Peachey.

Peachey led me to a small chamber behind the huge image tank that stood in the center of the room.

"The All Time is part of nonspace," Peachey continued. "Since nothing really goes anywhere in the holographic nonspace of the brain, time is, in reality, infinitely flexible. In nonspace, you can make a *moment* of experience as long or as short as you wish. You will see in the nonspace of Fullview that you will be able to manipulate *any* and *all* of the time in our two-hundred-million-year image archive."

There was just enough room for the two of us inside the small chamber. Peachey delicately ran her long fingers over the smooth, seamless, and highly iridescent walls of its inner surface.

"You can see, Betty Jean," Peachey said, "we're inside a small Fullview computer. This will be your *workstation* during the experiments. The dinosaur and other animal and human images we extract from your brain will be erected inside the larger Fullview computer. The brain images transmitted to Fullview make up the nonspace of Fullview. Fullview will display the holographic patterns, *holocopies* for short, of things going on in the nonspace of your mind—the dinosaurs, the sights, sounds, your homunculus, and your body. Your body's holocopy will be sent just as you see it in your memory. And in the larger Fullview, your holocopy will be *alive*, just as all of the images in your brain are alive! Do you remember our little discussion about this at the Nano-soft conference?"

"Yes, I remember," I replied. "All of the thoughts and images in our brains are alive because the neurons in our brain are alive. They are living holograms. But I will also be *alive* in Fullview, right? I will be alive in two places at once?"

"That's right," Peachey said. "You will be alive and thinking in both the small Fullview workstation and in the large Fullview at the same time. Tomorrow, Professor Vandervoort will explain to you that if you travel at the speed of light, you can be in two places at once! You will be traveling back and forth between your small Fullview workstation and the larger Fullview at the speed of light. You'll find this to be very simple the way Professor Vandervoort explains it."

"Wow," I exclaimed, "I can't wait to hear what Professor Vandervoort has to say about being in two places at once!"

"Yes, *wow* and *wow* again!" exclaimed Peachey. "You'll be in nonspace the whole time."

"Ah, nonspace," I replied. "Can I ask you about nonspace again, Dr. Peachey? Is it really real? I mean, I remember your discussion of nonspace at the Nano-soft conference, but how could there *really* be a *non* space? Wouldn't that just be *nothing*? Anyway, I thought Albert Einstein said that everything had to exist in a blended combination of space and time, the space-time. You're not contradicting Einstein, are you?"

"Einstein wasn't talking about space and time *in* the human mind," Peachey replied with a grin. "He wasn't talking about the human mind at all! Einstein was talking about everything *but* the human mind. What he discovered about space and time in the world *out there* does not apply to the nonspace of the All Time inside the brain."

"So how do we know about nonspace at all"? I asked.

"Nonspace is the reality of the entire world and the entire universe that you experience inside your skull," Peachey continued. "Nothing really goes anywhere inside your skull, you just think it does. Inside your brain, Mars is no farther away than the grocery store. And a trip to Mars takes no longer than an imaginary trip to the grocery store. *Space* is only needed if something is actually going somewhere—if it is going from some point A to some point B. Everything in the nonspace inside the brain is on a treadmill, so to speak. Professor Vandervoort pointed out that Einstein used only the nonspace reality inside his own brain when he mentally imagined what it would be like to travel alongside a beam of light, and that led to his most important breakthrough. So Einstein's space-time is actually a product of his own nonspace."

"It is so weird to realize that nothing really goes anywhere inside my skull," I said.

"Let me explain it this way," laughed Peachey. "Your brain operates a little like a CD player. Remember my little lecture in Redmond were I mentioned that Jill Price's super autobiographical memory was like a videotape of her entire life since she was about fourteen years old?"

"Yes, I remember that, Dr. Peachey," I said. "I was so amazed at Jill Price's super autobiographical memory that I looked her up on the Internet. And part of the reason I remember it so well is that one of my favorite actresses, Marilu Henner, is also one of the handful of people who have super autobiographical memory."

"That's right," Peachey went on. "All of your experience is like a gigantic, tens-of-millions-of-years-long movie that's recorded on a super holographic CD. In this *movie*, just as in real life, it appears that people and things are moving here and there through space. But there is no real *space* in a CD, only a changing pattern inside another pattern that really isn't space. The whole movie takes place in a nonspace. We have to come to realize that the entire universe can be represented either in space or in nonspace, and it doesn't make any difference which we use on an everyday basis. But the real, final universe that includes the only All Time human beings can know is in the nonspace of the human mind."

"So my brain acts as the holographic *CD* to produce the nonspace in the larger Fullview?" I asked.

"Correct," Peachey replied. "When we take the animal images from your brain, you will always be inside your own small Fullview workstation. But at the same time, you'll believe that you're inside the nonspace of the larger Fullview monitor with all of your animal images—just like you now believe you are in this room with me right now."

"But I *am* in this room with you right now," I said hesitantly.

"Not really, BJ," Peachey responded. "You are actually *in* your own brain—*all* of you, every molecule. And that's the only place you ever have been or ever will be. This room is in your brain and so am I. Oh sure, there is really a room *out here*, and a *me* is really *out here*, but your total experience of the room, me, and everything else is always *in* your brain. There is actually a tiny version of me inside your brain, and you're looking at it and talking to it. Next time you see a beautiful sunset, with clouds trailing off for miles toward the horizon, stop and think about what's really going on for a minute. The sunset, the clouds, and the beauty of it all are being recreated inside your brain. And that's the *only* way you can experience them—by seeing them as they are made *inside* your brain, as images. The brain creates a replica of the world, or at least of some of its features. This replica includes the illusion of the vast distances, the perspective, and the beauty of the sunset!"

Peachey caught the look of disbelief that was creeping across my face.

"I know this is a little difficult to understand," she said. "It's the most powerful illusion that the brain creates. When you look around this huge lab, it's hard to imagine that what you are seeing is crammed into a nonspace pattern *in* your brain. Look at it this way. What happens if your brain is damaged or diseased or if you take hallucinogenic drugs?"

"It changes the world for you, sometimes in pretty bizarre ways," I replied. "Anyone who has taken as many psychology courses as I have knows that."

"BJ, you would be surprised how many psychologists don't take that next step beyond all of their own course work," Peachey lamented with a sigh. "They never realize that the entire world is really the world inside the brain."

"Okay, I think I get what you're saying, Dr. Peachey," I said slowly, thinking carefully about what she had said. "But isn't it just as easy to think of the world as being *out there*, just as I have done all my life?"

"It's permissible to do that for everyday purposes," Peachey replied. "However, from a scientific standpoint, it's just not the way we humans really experience the world. If we want to understand nonspace and how it works, we must put the world, our bodies, and even our selves—in other words, everything—inside the brain where it actually is. After all, BJ, we are going to have to put this same *everything* into Living Windows 2016, if we expect it to think!"

"Yes, I see it now!" I exclaimed as the revelation of what Peachey was saying hit me. "The world I see *out there* is actually some sort of a hologram that my brain makes inside my skull."

"But there's even more to it," replied Peachey. "And it's even more bizarre. Your body is in your brain too. Yes, your body is real, just like the people and the computers in this room, but the body you know about is only in your brain. Research on *phantom limbs* has shown conclusively that your own body is some sort of a hologram, as you put it, that your brain manufactures. Always remember, your inner body, your homunculus, is the real you."

"Yes, I remember that from one of my psychology classes," I replied. "People who have been born without a certain limb or maybe lost a limb in an accident often feel the limb as if it were still attached. I think it was Ron Melzack and his team at McGill University that pointed out that these phantom limbs come from the brain—where they always had been, even in people born without a certain limb."

"That's right," Peachey went on, "all of the instructions that allow you to experience your entire body are, like everything else, inside your brain. And these instructions can make even a missing limb *exist* for you. As a matter of fact, BJ, we will not only be able to give Living Windows 2016 your thoughts, we will also give it the feel, the presence, and the motion of your body! We will give it phantom human limbs."

"Okay, so you're saying that my body is going to go into Fullview along with my thoughts, right?" I asked. "Good god, I just realized what you said. Living Windows 2016 will get my body too!"

"Your body goes too," Peachey said, grinning as if she had something up her sleeve. "And eventually, your body will be able to play in cyberspace. It's impossible to separate your body from your thoughts. You have to realize that, just like your awareness of those people standing over there, your awareness of your own body is generated in the nonspace of your own brain. You'll be able to interact with all of the animal images in any way you wish. But you'll have to keep in mind at all times that the images of your own body and your animals are coming from your own brain—and the brain of your *brain mate*."

Brain mate, I thought, *that must be the male with the perfect brain like me, the other experimental subject who was chosen.* Peachey could see the wheels turning in my head.

"By the way," she said, "your brain mate is the male subject you'll be ... uh ... traveling with. I am afraid you won't be able to meet him in person before you'll be meeting him in the nonspace of Fullview."

"Why aren't you going to let me meet him now, Professor Peachey?" I asked. "Are you afraid it will affect the quality of my dinosaurs?" I said jokingly.

"Actually, BJ, it's not that I won't let you meet him," Peachey replied matter-of-factly. "Your brain mate is up in Redmond, Washington, in another lab very similar to this one. His nonspace lab is on the Nano-soft campus. The Nano-soft software scientists wanted mirror-image control over this research. He'll appear here with you, and at the same time, you will appear in nonspace up in Redmond with him. I'm not sure whether you'll ever get to meet one another in person like you and I are right now. We'll see. But one thing is sure, you'll get to know one another in nonspace."

Peachey looked at me with a warm, friendly smirk, as if to see if I was getting some private joke. I didn't get it.

CHAPTER 8

Two Romantic Dinners at Once

(2014)

The next morning, I reported back to the Living Windows 2016 lab. This was the day I was to have the holographic image-gathering electrodes implanted in my brain. Peachey and two technicians were waiting for me. We went immediately to the fMRI lab, where my brain had been scanned two weeks earlier. They sat me down in a soft leather chair that looked a lot like a dental recliner. As my body sank into the warmth and smell of the leather, a small fMRI unit about the size of a beauty shop hair dryer was placed over my head.

I had long imagined that the surgery for the electrodes would be an ordeal. The first big surprise was that the electrodes were so small! Each electrode was about the size of a grain of sugar. I asked Peachey how such a small electrode could do the job. In her typical professorial style, she gave me a mini lecture on the topic. She explained that holographic images in the brain are distributed across web-like neural structures. Back in the 1990s, the original holographic image-gathering devices had been designed in Karl Pribram's brain lab at Radford University. Peachey told me they had been greatly improved by nanotechnology engineers at American Nonlinear Systems. The holographic image-gathering devices could now record holographic images of every mental event that goes on in the brain tissue around them.

The second big surprise, and relief, was that the actual placement of the grain-of-sugar-sized holographic image-gathering electrodes in my brain was rather anticlimactic. One of the technicians slowly and gently arched my head back.

"Open wide," he said.

While fully awake and sitting in the recliner with the small fMRI over my head, an automated electronic gun that looked like a computerized syringe began positioning itself gently against the roof of my mouth.

"I don't know where you got the idea that we were going to drill those holes in your head you mentioned yesterday," Peachey commented as the gun slowly positioned itself. "The roof of the mouth is the gateway to the brain, Betty Jean. A micro laser will open an extremely small channel into each one of your three brains, then the holographic image-gathering electrodes will be pneumatically puffed to the end of the channel. We'll cool the electrode channel slowly for about fifteen seconds while the entry channel collapses behind the electrodes. That's all there is to it."

After about twenty seconds, a technician who had been standing a few feet away and in front of a small Fullview monitor reported, "The reptile image-gatherer is sending."

I had felt only a small warm pressure. It felt about the same as gently pressing my tongue against the roof of my mouth. The gun was very slightly repositioned and again pressed gently against the roof of my mouth. Another twenty seconds or so passed.

"The PM image gatherer is sending," the technician reported.

"What's the PM electrode?" I asked Peachey.

Peachey smiled with a playful grin. "That electrode is in your mammal brain. *PM* stands for the paleomammalian brain. Paleomammalian means 'old mammal.' It is the brain of ancient mammals, the apes, dogs, and so forth—remember, Betty Jean? The PM brain was also quite dominant, we think, in early humans like *Homo erectus*. We're very interested in the *mental* images that will come out of erectus. Let me just say that the PM layer of our brain is of course the first layer that got *nipples* to deal with. That's what the PM brain is all about, nipples!"

Before I could motion to Peachey that I wanted to know what was so important about the brain getting "nipples to deal with," I felt the gun repositioning itself for the third time. It pressed gently. This time, a full minute must have gone by when the technician signaled that the Broca's speech area image-gatherer was sending. Broca's area, I knew, was the speech area of the human brain.

"She's successfully wired, Colonel Peachey," reported the technician handling the monitor.

The technician was all business. *Must be military*, I thought.

"Dr. Peachey, what's the joke with the *Homo erectus* brain and nipples?" I hesitantly inquired.

With a mischievous gleam in her eyes, Peachey glanced toward the technician at the monitor. "We think the entire civilized brain is run by nipples."

I looked at Peachey in astonishment at her remark.

"Just kidding!" Peachey laughed while she put her arm around the technician and looked playfully at me. "*Nipples* is just a simple way for us to refer to the PM brain machinery, Betty Jean. Mammal, mammary glands, nipples—get it? With the evolution of mammals and their nipples came a brain that nursed and cared for its young. Nursing and caring are the beginnings of justice, that another person is as important as we are, that one person is equal to another. We're thinking the brain's PM images supply the basic images for *everything* civilized. Nipples are responsible for the development of all *law and order* down through history. In fact, without nipples, there would be no democracy!"

"I had no idea," I said in an amazed grin.

"There's more," Peachey continued with a smile. "We now know that the nursing and caring that developed with the PM brain leads to belief and conviction—to care is to believe. We are vitally interested in how those images are related to the law and order of science and mathematics. Without belief and conviction, there would be no scientific theories. Mathematics would be just a bunch of meaningless patterns of gibberish, so understanding the images of belief is vital to getting Living Windows 2016 to believe. It must believe so that it can think."

Peachey suddenly glanced at her wristwatch.

"More later," she said quickly. "We are scheduled to meet with Professor Vandervoort, and we definitely don't want to keep *him* waiting. Betty Jean, Professor Vandervoort will be filling you in about what to expect when we begin extracting images from you and your brain mate tomorrow."

Off came the fMRI and the electrode gun, and I was out of the chair.

"You haven't seen my luxurious office," Peachey said while motioning me to follow her.

Peachy led me up a flight of stairs to a loft that overlooked her Nano-soft Living Windows lab. We entered a living room–sized area. Inside the room were several folding chairs sitting every which way and nothing else. The walls had the same smooth, highly iridescent finish that I had seen in the Samsing Seamless two-way retina and my Fullview workstation. In one of the chairs that was facing away from us sat an older man.

"Please sit anywhere," Peachey said to me. "Betty Jean, this is Professor Vandervoort."

The man remained facing away from us and said nothing. A full minute must have gone by. He remained motionless and silent.

Peachey smiled at me, nodded, and putting her head close to mine, whispered, "He's thinking, just hang on a minute."

"Betty Jean McKinney," the man finally said without turning, "I know you better than you do. But there's one thing I don't know about you. I don't know how you will deal with a *paradox*."

He went silent again.

"Oh, you're wanting me to turn around," he said after a long moment. "Well, I don't care to turn around right now."

What a strange duck, I thought.

"I don't want you to *look* at me," he went on, "I want you to *listen* to me and listen carefully. Can you do that?"

Somewhat caught off guard but intrigued, I reflexively replied, "Hit it, Doc!"

I sheepishly looked at Dr. Peachey and shrugged, as if I'd been a little too forward. She smiled and shrugged back at me as if to say "What the hell, go for it!"

"I will call you BJ," he began with a tone that inferred that's what he'd call me whether I liked it or not. "BJ, what's a *paradox*?"

To my surprise, his abrupt manner told me I could do nothing but obediently try to answer. I wanted to tell Professor Vandervoort all about the talking boxes and the paradoxes their fake worlds got them into, but at the same time, I wanted to see where this strange man would go with his question. So I decided to act a little dumb.

"A paradox is a contradiction, right?" I replied without hesitation. "I do remember that paradoxes were discussed in a philosophy class I once had, but I can't put my finger on any of the details."

"Okay, okay, the liar paradox," Vandervoort interrupted, already sounding slightly frustrated with me. "You're right, a paradox *is* a contradiction. Let me refresh you about the contradiction in the liar paradox."

Vandervoort scribbled something on a piece of paper.

"Read this sentence," he said, passing what he had written back over his shoulder.

The note simply said "This sentence is false."

"Well," Vandervoort said impatiently.

"Please let me look at this for a minute," I said.

I had seen the liar paradox in my philosophy class, but I pretended to have a sudden insight.

"Oh, I see, Doc," I said, "if the sentence is *true*, it is *false* at the same time, Doc. Because it is saying it is false, it is true. The sentence is both true *and* false, so it contradicts itself. It's a paradox."

"Not bad, not bad," Vandervoort said, still facing away and nodding his head. "The *liar* paradox is a boring little piece of trivia, but you have done well, BJ. This version of the liar paradox is a stupid little game, which does nothing but make university philosophy professors think they sound smart. But the existence of paradox *itself* is probably the most important discovery made in the entire history of mankind. I will not give you the details now, but you will discover them during the journeys you and your brain mate will begin tomorrow. Just let me leave you with this thought for now. Paradoxes show us that the raw power of the human mind can outthink even its own language—as you just did with this little sentence. Only on those rare occasions when the mind outthinks the contradictions in its own language, like when it sees its way through a paradox, does it have a chance to glimpse itself in its pure form. In its pure form, the human mind can even outthink language. BJ, during your journeys, you and your brain mate are going to get more than just a glimpse of the pure, undistorted human mind. Your pure minds are waiting deep in the ancient layers of your brains and will be released to dominate every word you'll ever utter or ever hear. And we are going to put copies of that deep, pure mind into Living Windows 2016."

Professor Vandervoort finally turned around to face Peachey and me. My god, I didn't know what to think. He looked so much like Sean Connery that I had seen in old 007 movies I could hardly believe my eyes! Although he must have noticed my surprise, he remained absolutely expressionless. He moved to a chair closer to mine.

"Now," he said, raising his finger to emphasize the point, "I want to see if you can catch glimpses of your true mind as you try to outthink a fresh, new paradox that no one has ever heard. I was thinking this one up as you and Dr. Peachey came into the room."

"I must lay out some background for you, BJ, but it might get a little technical, so bear with me," he began. "BJ, how would you like to have two romantic dinners with two different men in two different cities *at the same time*? And neither of the men would know about the other."

"I think I'd love it," I replied enthusiastically, wanting to play along. "But that sounds pretty impossible."

"Maybe not," Vandervoort went on. "Did you know that if you could travel at the speed of light, you could zip up to Nano-soft in Redmond and then back here to San Jose more than one hundred times in one second?"

"Let's see now," I thought out loud, "light travels at about 186,000 miles a second, and it would be about 1,400 miles for the round trip up to Redmond and back. Yes, I think you're right, I *could* make the round trip one hundred times a second. Actually, I think it would be slightly more than one hundred times a second. But I get your point, Doc, I'd be up to Redmond and back one hundred times a second."

"Well," Vandervoort said with a Connery-style smile on his face, "because flicker fusion is only sixty times a second, far slower that one hundred times a second, you could be in both San Jose and Redmond at the same time. Both your mind and your body would be physically present at two different romantic dinners at the same time."

"Wait, wait, wait," I interrupted, "I don't get it—what's *flicker fusion*?"

"Oh, that's simple," he replied. "It's the same thing that happens when you watch a movie, television, or even those fluorescent lights up on the ceiling. They're all actually going on and off—flickering—at a high enough rate so that the flickering disappears. All we see is a steady picture or a steady light, *and we can't tell the difference.* The brain thinks that *anything* that flickers at least sixty times a second is present all the time. It's the brain's reality, and it can't escape it. So both the man at the dinner table in San Jose and the one in Redmond would think you were there all the time and absolutely real. There's no way that either of them could tell that you were really pulsing on and off a hundred times a second. Both of the men could both talk to you and reach out and touch you."

"I'd be pulsing in front of both of them at one hundred times a second," I said, "and that's much faster than the sixty-times-a-second speed, where things become visually and physically present—flicker fusion, you call it. So I would be in two different places at the same time, right?"

"Right!" Vandervoort continued as he scooted his chair closer to me. "Now, here's the really interesting thing. To your brain, the two men flickering on and off one hundred times a second would also fuse into solid people for you to see and touch. Your brain would create two simultaneous realities, one in San Jose and another in Redmond. Einstein proved that even if you travel at the speed of light, everything around you appears to you to be normal. This was shown in Star Trek as the starship Enterprise went faster than the speed of light; everything inside the starship appeared to be normal to the crew. Just as the two men could talk to you, you could talk to both of them. They could touch you, and you could touch them. You could simultaneously consume the two dinners. After your romantic dinners, you could even fall in love and make love to the two different men at the same time."

"Oh, that would be interesting," I blurted out.

"Yes, as a matter of fact, it would," Vandervoort said with a slightly dubious smile. "Because you'd know you were doing it. In your mind, you would be like a person who has two secret lovers. You'd know all about your two lovers, but instead of spending separate times with each one of them, you would be with both of them at the same time—they would be completely separate but simultaneous lovers. We don't know exactly what your experience would be like, but perhaps it would be something like riding a bicycle and juggling at the same time."

Aha, I thought to myself, *flicker dining, flicker romancing, flicker lovemaking—these guys are weird, weird, weird.* The whole slant on Vandervoort's paradox reminded me of Peachey and her *time sled*. I was beginning to think that Freud was right. All of this science stuff is nothing more than disguised sex, but I had to admit, the idea of flicker lovemaking *did* have an appeal.

"BJ, begin to think about the two flickering people you have become," Vandervoort said, jolting me from my flickering fantasies. "Think about how your brain would be living two separate lives instead of only one. Think broadly—perhaps if, since childhood, you had lived your whole life flickering, that perhaps you would have become an attorney in San Jose and a doctor in Redmond. Can you imagine that?"

"Yes," I said, "I suppose that's really possible."

"Okay," Vandervoort said, "you're ready for the contradiction—you're ready for the paradox."

All this time, I had been thinking that just being in two places at once was already a paradox.

"What would happen if you were shot and killed while having dinner in San Jose?" Vandervoort continued. "Would you die, or would it be possible for you to be just *half*-dead? Remember the sentence in the liar paradox that was both true and false? In a similar way, could you be both dead and alive?"

My first reaction to Vandervoort's new paradox was that I didn't even know how to think about it. Vandervoort stood up and began walking the room.

"Well," Peachey piped up, "I—"

"Please, Edythe," Vandervoort interrupted, "let her think, let her ponder this for a moment."

Peachey obediently shut up, putting her finger to her lips and nodding to Vandervoort. Vandervoort continued to pace the room. It was amazing, as his pacing gobbled up the distance around the room, this guy even had the panther-like walk of Connery. So stunning was his movement I momentarily forgot I was supposed to be trying to figure out whether I was dead in San Jose or alive in Redmond or both!

Vandervoort again awakened me from my fantasies. "BJ, I have dragged you through the paradox of being in two places at once for a reason. I wanted to begin to prepare for your trips back to the ancient parts of your brain."

"Betty Jean, there is only one way we can send you back to the land of the dinosaurs and ancient mammals," Peachey chimed in. "*We will flicker you back at one hundred times a second.* The electrodes we implanted in your dinosaur brain will flicker your dinosaur images into form. At the same time, the electrode images from your human mind and your homunculus will flicker back to meet them. Your mind and inner body will be both in the land of the dinosaurs and also here in the lab in your Fullview workstation at the same time."

"Please, Dr. Peachey, let me finish explaining the situation," Vandervoort again interrupted.

"BJ," he went on, "just as you were able to reach out and touch your two lovers in the two-places-at-once paradox, you'll be able to reach out and touch your dinosaurs. Everyone has seen dinosaurs, but the dinosaur part of our brains has never seen a human. These creatures have been locked away in our brains for tens of millions of years."

Vandervoort stopped abruptly and hesitantly glanced at Peachey.

Peachey nodded. "Go ahead, Professor Vandervoort, we *must* tell her before the experiments begin."

I didn't like the tone of Peachey's voice.

"What must you tell me?" I timidly asked.

"We have a small problem with this part of the experiment," Vandervoort replied.

Oh, great, I thought to myself, *I hope he's not going to tell me that I might end up in two places at once, half of me in Fullview and half of me in who-knows-where.*

"We are not sure what will happen to you if something unexpected goes wrong while you're in the land of the dinosaurs," Vandervoort went on. "Just as you can reach out and touch them, they can reach out and *touch* you. We don't know what would happen if you were attacked by one of your dinosaurs. These creatures from the depths of your own brain may find you to be a tasty-looking little morsel—they may try to *eat* you. Remember, even though they will be coming from your own brain, your own dinosaurs have never seen a human before. I don't wish to unnecessarily frighten you, but the experience of being eaten alive, even by holocopies of your own dinosaurs, might shock your human mind into a coma, or it might drive you permanently insane. We just don't know what the effect would be."

Vandervoort leaned over close to Peachey and me, putting one hand on my shoulder and the other on Peachey's.

He whispered playfully while quickly looking back and forth at us, "But there's a theory going around that while being eaten, a primitive death instinct would take over, *and you might actually enjoy it*! You wouldn't really be being eaten, you'd just experience being eaten. If that happens, you should simply go limp and try to enjoy it."

"*Very funny,*" I replied, politely going along with Vandervoort's little joke. "Being eaten by creatures coming to life in my own brain, isn't that a little far-fetched?"

"Oh, no," interrupted Peachey. "Ancient parts of the brain actually terrorize us constantly. That's why we're concerned about them. Every time you're afraid, feel like you would like to kill someone, or get jealous, the creatures of your ancient brains are attacking the civilized thoughts of your human brain. You just don't see the creatures that are doing it, but they're there. Sigmund Freud called them *creatures of the id*, of the deep, instinctual unconscious. He thought that they caused all of the horrors we humans experience. According to him, they are the source of all devils and of all hell."

"Okay, okay, maybe there could be a problem of being attacked by the dinosaurs of my own mind," I admitted. "But I certainly wouldn't *enjoy* being eaten!"

"Wrong again, I think," Vandervoort said. "Freud *also* thought that these ancient creatures provide our brains with a death instinct. It is called Thanatos. Freud didn't work out the details of the death instinct, but he thought that we are constantly driven to return to a prelife form—to die. The death instinct is the most powerful human drive, even stronger than sex. And the potential pleasures of the death instinct may be far greater than sex. Buddhists call the return to prelife "nirvana" the ultimate experience. If the dinosaurs go after you with a gleam in their eyes, you may have the biggest opportunity any human being can ever have. Survive it, and you will want to do it over and over again. But if you succumb to it, you may become permanently insane!"

Peachey stood up, raising one eyebrow to Vandervoort. "Whoa there, big fella. Betty Jean, everything Professor Vandervoort is saying is true, but we really don't know if you'll even be attacked, let alone be eaten. You will always know that you are in two places at once. As far as being eaten is concerned, we think you'll be fine—*if*..."

Peachey abruptly broke off the subject. *If what?* I wondered.

Peachey quickly changed the subject. "Betty Jean, let me give you the details of how we are planning to use all of these experiences you're going to have. I think you should know how this project might change *everything* human beings do and think *forever*. As I have told you, the key to the entire project is to obtain software for Living Windows 2016, software that works the same way the human brain does. As the holographic image-gathering electrodes in your brain harvest the animal and human images that make up your mind, we will be copying the details of the whole thing to electronic files. We will have, for the first time ever, software that thinks and feels just like human beings do, instincts and all. You'll not only be you, Betty Jean, but you'll be an *abstract* human mind on a disk. This abstract, electronic version of your mind will be Living Windows 2016. God, I love the sound of that—Living Windows, Living Windows!"

"I don't understand how I can be an abstract mind on a disk," I sheepishly admitted. "Will I know I'm on a disk?"

"Colonel Peachey, let me handle this one," Vandervoort interjected. "This is my favorite part of the *new world* we are creating. BJ, this may come as a shock to you, but your human mind is *already* 'on a disk,' so to speak. Your brain is a gelatinous nonspace *disk*."

Vandervoort's explanation reminded me of what Peachey had said about how nonspace worked. There is no *space* inside our skulls. Our real world of experiences occurs in a nonspace.

"That means," Vandervoort continued, "when we have all of your animal and human images on the disk, *you* will just be in a duplicate nonspace. Your All Time, your two hundred million years of immortality, will be occupying two nonspaces, one in your brain and another one on a disk. As Peachey has probably told you, your inner body will also be on the disk. So you won't feel like you're on a disk—you don't feel like you're in a *brain* now, do you? You'll feel like you're in a body, just like you do right now. You will be in a whole new *super* world of journeys with which you can directly interact anytime you wish. You'll be able to go to the land of dinosaurs, the land of the ancient mammals, or sail through cyberspace, body and all. You will not live in the *present* like the rest of us, but you will be able to interact with us if you wish. You will live in the All Time, which extends from the first dinosaurs all the way to the ever changing *now*, the *present*. Because you will be able to visit any time along the past of the species, you will, from that time leading forward, have a *memory of the future*. Think of it, BJ, you will not only be able to remember yesterdays but the tomorrows you have already known as well! But there will always be another unknown tomorrow as the *now* moves forward, and if all goes well with our experiments, everyone will be able to join you in these journeys through the All Time nonspace."

"The All Time," I said as I pondered the idea. "Wow, what will that be like? I know what it is, but it sounds unbelievable."

"Ah, *time* itself is a funny thing, BJ," Vandervoort replied in a philosophical tone. "The *now* for animals is very narrow. They have a tiny bit of the past and virtually no future in their *now*. As animals evolved, their *nows* got bigger and included more and more of the past, and slowly, they began to get a little of the future when they planned ahead. Humans have the biggest *nows*. In our *nows*, we can think about all kinds of past events and all kinds of future events. And the most intelligent human beings have the biggest *nows* of all. In fact, it's their big *nows* that make them the most intelligent."

"Are the human *nows* really measurably bigger than the animals' *nows*, Professor Vandervoort?" I asked. "I mean, are *nows* scientifically *real*?"

"Oh, yes," he replied with a laugh. "The animal *nows* of dogs and cats are only two or three seconds long. Chimpanzees' *nows* might be about seven or eight seconds long, but in humans, the *now* has expanded to twenty to thirty seconds. Just think, BJ, when you're on a disk, your *now*, the All Time, will be over two hundred million years long. And because the images of the All Time of nonspace are already in your brain, you can run them at any speed you wish. You can run a thousand years in ten seconds if you wish, or if you wish, you can have a *now* that's a thousand years long. You will be able to grasp the meaning of everything that happens in a thousand years in one thought, one conceptual grasp! Can you imagine what that will be like, how intelligent you would be? You could be the greatest *prophet* who ever lived!"

"But I will still be *me*, won't I?" I asked hopefully.

"You'll still be you," Vandervoort replied with a smile. "There will be two of you, you and the other you on the disk of Fullview in the All Time of nonspace."

"Will this *me* know the other me who will be in the All Time?" I asked. "God, the other me will be so intelligent! She'll be like a god."

"Yes and yes," Vandervoort replied. "One of you will be living in a world where there are paradoxes, the other will have escaped paradoxes forever. You will know the other you whenever you are constructed in Fullview."

"Let me amplify that. Yes and yes, Betty Jean, the other you, who will be able to cruise the All Time at the speed of light, will be like a god, and you will still be able to talk to her," Peachey said, entering the conversation. "You will begin to understand both of these things tomorrow morning when you enter the All Time of your nonspace."

CHAPTER 9

The Journey 1: Into Nonspace and the Joy of Being Eaten

(December 1, 2014)

December 1, 2014—this was the most exciting and important day of my life, more important than being conceived or even being born. This was the day that I began my journey back through everything that had been packed into my DNA and into my brain over the last two hundred million years. I knew it was silly of me, but last night, I had asked Dr. Peachey what I should wear for my trip back to the land of the dinosaurs. She told me not to worry about it because my mind would dress me as it saw fit. *Boy*, I thought, *I wish it could do that all the time*. But little did I know about how my mind would like to *dress* me.

How I looked was important to me because seated around the huge Fullview monitor were dozens of scientists and technicians. Professor Vandervoort, of course, was there, but there were technicians, botanists, paleontologists, psychologists, neurologists, and who-knows-who. Peachey had told me that Pauline, the blind girl who started all this, would be there. As I approached my Fullview workstation, I looked over toward the large Fullview monitor for Pauline. I quickly spotted her wearing her tiny video camera that fed images directly into her brain and sitting right next to Professor Vandervoort. I was about to be "onstage" for my greatest performance ever.

As Peachey and I approached my Fullview workstation to prepare me to begin my first journey, a question that had been haunting me all last night suddenly became urgent again.

"Dr. Peachey," I said, "I was bothered all last night by something you said or, rather, something you didn't say yesterday."

"What's that, BJ?" she asked.

"We were talking about the possibility that my dinosaurs might attack me," I continued. "You said an attack would be all right '*if* . . . ' and then you quickly changed the subject."

Peachey's face took on a rare seriousness.

"I must be absolutely frank with you, BJ," she said. "We *think* you'll be fine. But we're not 100 percent sure. We *know* you'll be fine *if* you can *always* remember that everything you see and feel is coming from your own brain, but if you lose your focus and you begin to believe that your dinosaurs are really real, your mind could become imprisoned in your own dinosaur land forever. Your body would still be alive here in the lab, but we could never get the mental part of you back here to 2014. I know you were surprised when Professor Vandervoort told you that being torn apart and eaten by the dinosaurs might be so tempting that you'd let it happen. The problem is that you might suddenly be in trouble as your body is being torn apart. You might think you are beginning to die, and you may not be able, at that point, to pull your senses back together—in fact, *you may not want to*! We just don't know what will happen."

"I can't believe Professor Vandervoort would think such a strange thing, Dr. Peachey," I said. "Why would I *not* want to keep my focus if I were being killed?"

"Well, as Professor Vandervoort told you yesterday," Peachey answered, "it is the dinosaurs in people's minds that cause all of the mental illnesses. When the ancient part of the brain gets the upper hand, the *human* portion of the mind can't handle the pure animal forces in it, and it goes *crazy*. The human portion can become imprisoned in the holographic records of the dinosaur brain, the same holographic records Jeffrey Dahmer was forced to act out. And this could be coupled with the death instinct. It all could suddenly hit you on a massive scale."

I'll show you, Dr. Peachey, I thought to myself.

"You don't have to worry about me, Dr. Peachey," I said, "I'll show you and Professor Vandervoort that the idea of the sublime beauty of the *death instinct* theory is just a bunch of bull."

Peachey gave me her raised-eyebrow look as she popped her head back in approval. "I'm sure if anyone can, you can," she replied as she began to leave my Fullview workstation. "I'll be observing you in your dinosaur land, and I'll be on the lookout for trouble."

With a wink, she closed the glass door to my workstation, and I lay back on the electrode-sensing headrest. *Here I come, my dinosaurs*, I said to myself as I commanded Fullview to "Initiate Journey 1."

"Look at her over there in her Fullview workstation," Peachey said to Vandervoort and a collection of scientists and technicians seated around the huge nonspace 3-D monitor. "I don't think she realizes how alone she is. Yes, she's here in the lab, but she's also all alone, headed two hundred million years back in the depths of her own brain. God, I hope she can make this first journey okay. Damn it, if there's going to be trouble, this first journey is where it will show up."

As if awakening from a good night's sleep, I suddenly found myself walking somewhere that seemed right out of the ancient Jurassic era of the dinosaurs.

"Look, *there* she is, 'walking' in nonspace," announced Vandervoort as he stood and pointed toward the huge Fullview monitor.

"Hey, you guys," I yelled, "I have just entered some sort of fantastic primordial jungle. Can you see what I am seeing?"

"Yes, we can see you walking in nonspace, taking in the scenery," Peachey replied.

I could not believe how beautiful and huge these ancient plants were. I had never seen so many colors before. And they were so bright! Maybe everything was so vivid because deep parts of my own brain were producing all of this, I thought. I timidly reached out to touch a large cone-shaped bluish plant that seemed to be vibrating. To my surprise, it was extremely soft and rubbery.

"I'm not quite sure I totally get this nonspace idea," one of the observing scientists called out. "How can Betty Jean be meeting her dinosaurs in *no* space? What does *nonspace* really mean?"

"It's actually quite simple," Vandervoort replied loud enough so that everyone could hear. "The ancient dinosaurs stored in our genetic memories, as big as they are, don't take up any *space* at all in our brains. They're just holographic information patterns that your brain interprets as taking up space. Look around the lab, everyone. It looks huge, doesn't it? But inside your brain, where you actually *see* it, it's what we call *nonspace*. It takes up no room at all. *Everything* inside the brain is manufactured in nonspace. I love to go to the beach and look out upon the immensity of the ocean as it rolls in huge waves. Looking at those ocean waves coming in, I often realize that inside my brain there is no *immensity*, no *huge waves*—just patterns upon patterns, none of which take up any measurable space. So when we are *in* our minds, as BJ now is, we're in nonspace. BJ thinks she is walking right now. But inside her brain, everything is on a virtual treadmill, so to speak. Nothing actually goes anywhere when we are imagining things in our minds. They only seem to. Think about this, you scientists. If we can crowd an *infinite* amount of *stuff* into our brains over a lifetime—and most brain scientists believe that we can—then none of it could be taking up any space, could it?"

Vandervoort's impromptu little audience was silent for a moment. As the idea of nonspace slowly sank in, everyone began nodding in thoughtful agreement.

"People have a hard time with the idea of nonspace, don't they?" Peachey leaned over and whispered to Vandervoort.

"Good cri-*min*-y, where is all the botanical scenery coming from?" asked a startled technician.

"It's coming from the ancient part of BJ's brain, just like the dinosaurs will," replied Peachey.

"This is incredible," one of the scientists loudly interrupted. "I'm a botanist, and we've only seen these plants in fossil form. We could never have imagined the coloration of that foliage. Those are the most beautiful living colors I have ever seen."

"Don't forget," Peachey piped up, "BJ is seeing those plants too. Look at her examining them, feeling them and smelling them."

"I'll bet you'd like to hop into a Fullview and go along with her right now, wouldn't you?" Vandervoort laughed back to the botanist.

Pauline, who had been sitting quietly with her miniature video camera tracking directly on Betty Jean's every movement, snapped her focus over to Vandervoort.

"I would like to go with her too," said Pauline in a wishful sigh. "It's so beautiful in there. I mean it's so beautiful in that part of Betty Jean's mind. Is it like that in my mind too, Dr. Vandervoort?"

"As a matter of fact, Pauline," Vandervoort went on, "even though you are blind, it might even be more beautiful deep in your mind, and if BJ is successful in her journeys, before too long, we may be able to send you or anyone else who wishes back to their land of the dinosaurs—the ultimate amusement park in their own minds. People could even go in groups."

"I will go there someday," Pauline responded with a tone of confidence.

"Dr. Peachey," I said, "I can see what you guys are seeing too. I mean, I can see *me* walking in the jungle, but I can also look around and see the jungle as if I were actually walking *in* it. I can switch back and forth between seeing me in the jungle and actually being in it just by choosing the one I wish."

"Don't forget, Betty Jean," Peachey explained, "as long as the Fullview computer is on with your electrodes activated, you'll be in two places at once. You'll be in your Fullview workstation, watching yourself, and you'll also be over here in the large Fullview monitor experiencing your ancient memory. Remember Professor Vandervoort's two romantic dinners at once? You're experiencing everything just as you should—you and the system are working together perfectly. And all of us out here are enjoying the same Samsing two-way retina. We can see you traveling deep into the layers of your brain, and you can see yourself and see us watching you. Remember when you met the army lieutenant Ann at American Nonlinear and the nanocameras embedded in the two-way retina? You can thank Samsing and those nanocameras for your godlike sentience."

"Hey, look!" interrupted the botanist. "The plants are being disseminated."

"No, wait," he went on, "they're growing back. What's happening? Good god, they're growing fast. It looks like one of those time-lapse videos—the plants are growing back right before our eyes."

"BJ's ancient dinosaur brain is decoding part of its own eating cycles," Vandervoort explained. "The dinosaur brain was a very close evolutionary spin-off of the laws of plant and animal creation. Plant growth was as basic to them as language is to us. Their brains lived in a world where whole jungles would be ravishingly eaten by them in one day, only to grow back nearly as fast. That brain is still part of us. That is why we, even as children, are fascinated by plant growth."

"Yes," added the botanist, "those ancient plants we're looking at were like some forms of today's bamboo. The stuff probably could grow a foot or two a day. Just think, everyone, an acre of that lush stuff growing at that rate could produce fifteen thousand bushels of plant food a day. What a farm! Do you have any idea how many cows you could feed with an acre of that stuff?"

"That's right," said Vandervoort. "The dinosaur age could have been called the age of eating. With the high levels of oxygen in the atmosphere in those ancient times, every living thing was on a rampage of growing and eating. That's why everything got so huge—*eat, eat, eat*. The dinosaur's motto was 'If it smells or moves, *eat* it.' Deep in the dinosaur part of our brain, there are images that recorded the millions of years of this ravenous eating. Those ancient brain images recorded a condensed version of what life was all about during the dinosaurs' existence. They are *eating* archetypes—in other words, ancient *image* patterns or templates in the brain that drive eating behavior. This is why we are able to see the plants and their fantastically rapid growth so clearly—and soon, I expect, we'll be seeing the dinosaurs."

"Look, Betty Jean suddenly looks frightened," Peachey said. "She appears to be frozen in her tracks."

"I think it has dawned on her that she's alone in a pretty strange and scary world," said Vandervoort. "Seeing all those plants suddenly disseminated and then quickly grow back has probably unnerved her."

I did feel frightened. Things were rustling and moving, and I realized I was in a *real* world of some sort. It was beautiful, but it was getting a little scary. And I was all by myself.

Then the strangest thing unexpectedly happened!

"Hey, where the holy hell did he come from?" shouted Headley. "And who the hell is he? How could he have gotten into the nonspace? I thought we had total electronic security on that Fullview computer. Is that a fucking hacker, trying to fuck up our 2016 project?"

"Ah!" Peachy exclaimed. "Professor Vandervoort and I have been expecting him. Betty Jean's brain mate has shown up."

"What the fuck's a brain mate?" demanded Headley in a fuming but curiously protective tone.

"Is that the only word you know, Headley?" barked an agitated Peachy. "Just calm down please, and I'll try to explain. That *man* you see was *created*, like everything else in nonspace, by Betty Jean's mind."

Peachey put her hand on Vandervoort's shoulder and leaned down near his ear. "Professor Vandervoort," Peachey whispered, "do you know why Headley suddenly has such a limited vocabulary? I don't like his tone. Perhaps he should be watched."

"Yes, I know," Vandervoort whispered back. "That's classic Jay Headley. Headley is a little difficult, but I think his attitude is harmless. He's my best Fullview electronics troubleshooter. I brought him in, and I should have warned you about what happens to him under stress. I know your other computer people are good, but we just may need Headley if anything bizarre goes wrong with nonspace. Besides, I think Headley has a crush on Betty Jean, and he probably doesn't like this brain mate stuff."

"But who *is* that hunk in there with Betty Jean?" asked one of the visiting female scientists. "How on earth did she come up with him? I'd like to have one of those!"

"Trust me, my dear, you do have one of *those*," Peachey piped up. "If you're lucky, real lucky, you'll be able to get your hands on it—him—someday."

I was wondering who this man was myself. Was he up in Redmond in the other Fullview that was synchronized with mine? Were we in this together? Was I being seen up in Redmond just as he was being seen here in San Jose? He looked familiar in a strange way. What was I supposed to call him? Why didn't Dr. Peachey introduce us *before* the experiment started? Just as Headley was ranting and raving, *I* too wondered what the *fuck* was going on. I was surprised that word popped into my mind just then. But it did.

"When Betty Jean became afraid, the electrodes in her brain began to pick up what we call the *brain mate*," Peachey replied, looking around to everyone. "This will get kind of complicated to those of you who are not psychologists. This may be hard to swallow for some of you. But here goes. Each of us is actually *both* female and male, and all of us are born with ancient female and male images in our brains. These images are the *models* of which of the two sexes our psychological worlds will become. If we are female, we usually *become* the female image as we grow up. If we are male, we become the male images. Little girls just naturally respond to the world in little-girl ways, and little boys in little-boy ways. The male images in the female brain and the female images in the male brain are our brain mates. These images of the opposite sex allow us to naturally accept and be driven by the fact that we must eventually mate and procreate during our lifetimes. But the image of the opposite sex also determines what we will look for in the ideal mate. We strive all throughout our lives to find the mate who fits the model of the perfect mate in our brains. We all keep looking, but few of us are lucky enough to find the perfect match."

"Yes," Vandervoort interjected as he again turned to address the little audience, "have you ever noticed how some people, of both sexes, just seem right to you and others don't? Part of the feelings of *love at first sight* are released by their close matchup with the brain mate image we carry in our brains. And the very fact that women are naturally attracted to men and vice versa is produced by the image of the opposite sex we are born with. Dr. Peachey, don't some of you psychologists call these brain mates the animus-anima complex?"

"That's right, some of us do," replied Peachey. "Animus is the male brain mate in the female brain, and anima is the female brain mate in the male brain. So the man who has just appeared in nonspace is Betty Jean's animus. He is Betty Jean's perfect male soul mate. Her brain produced him in Fullview because she was afraid. Actually, anytime any of us are afraid, our brain brings up the image of our brain mate to help protect us. We normally don't *see* the animus and anima because they operate at an unconscious level, but because Betty Jean's brain is implanted with holographic image-gathering electrodes, the Fullview computer produced an actual image for all of us, including Betty Jean, to see. I can assure you, folks, that we expected this to happen. Betty Jean's man will come to support her not only when she's afraid but whenever she gets mad or amorous."

"Dr. Peachey," said the botanist in an incredulous tone, "do you really mean to say that there is an *actual* image in each of our brains of this perfect brain mate?"

"Yes, absolutely," replied Peachey. "It's an actual image. In recent research on phantom limbs, Canadian researchers have shown that a *complete* image of our body is wired into our brains before birth. If we lose part of our body through an accident, or even if we are born without an arm or a leg, we still can experience that missing part. The missing parts are experienced as *phantom* limbs. In the same way, our brain mates are wired into our brains before birth. And we unconsciously live our whole lives with them. We see little glimpses of them in the people of the opposite sex that we like and, especially, the ones we fall in love with. Our brain mate is wired into our brains for the same reason the growth of vegetation is wired into the image archive of the dinosaur brain. The sexual brain mate is more important to humans than in any other species. It is only us humans that fall in love through a brain mate."

"There is one more thing—well, a couple more things," added Vandervoort. "First, it's the brain mate in our unconscious that gives rise to all of our superheroes. We like Superman because he is the perfect brain mate for all of us, whether we are males or females. He is invincible and wants to do nothing but protect us from evil. That brings me to the other thing about brain mates. Because we all have both male and female images deep in our brains, some people find that their perfect brain mate turns out to be the same sex they are."

Wow, I thought, this *is my soul mate, my brain mate? This is my perfect male, and he'll show up whenever I need him? Then I am not alone in this jungle. I have never been alone! I have never been alone in my whole life. He's been with me ever since I was born. God, what a new way to think about reality!* Anybody who was willing to get themselves *drilled* for image-gathering electrodes and get themselves into a Fullview could actually see their perfect brain mate.

"Dr. Peachey, I heard the explanations you and Professor Vandervoort gave," I said. "This is wonderful! Why didn't you tell me about this . . . this guy?"

"Betty Jean, I distinctly recall telling you that you'd meet your brain mate in the nonspace of Fullview," Peachey replied. "I just didn't tell you how he'd get there. I knew you'd have to see it to believe it."

"Can this guy talk?" I said.

"You'll have to talk to him first," Peachey said. "Then you'll have started a discussion with your *second self,* and away the two of you will go. Actually, he'll do *anything* your perfect male would do. He is *your* animus! He's literally your knight in shining armor."

"Dr. Peachey," asked one of the female technicians, "why is Betty Jean's brain mate wearing an army uniform like yours?"

"That's the way Betty Jean's animus complex wants it right now," Peachey replied. "Right now, she is flavoring him with part of her own female sexuality. She's momentarily confused about her own *natural* bisexuality. Her brain is mixing male and female images. She is comfortable with me, and so she's dressing him with *my* army clothes. It may mean too that she sees me as being sexually attractive. But as her journey moves on, she will likely change what her brain mate wears. We'll have to wait and see. Just remember, we are observing the very basics of her sexuality. How it will unfold is unknown to us. It will likely be a surprise to her too."

"Listen," said Professor Vandervoort. "Betty Jean is talking to him."

This is going to be interesting, I said to myself as I began talking to this handsome creature that stood examining every inch of my body in a sort of awestruck surprise.

"What is your name?" I timidly said.

Here I was talking to a man created by my own brain, but I felt somehow timid, as if I were on a blind date.

"My name is Larry," he replied. "I have been looking for you, it seems, like, forever. Are you all right?"

"Yes, I'm all right now that you're here," I said. "This jungle is so . . . so strange. It was kind of freaking me out. And I have this really strong sense that I am being watched by something."

"Don't worry," Larry said. "Remember, you're in the nonspace of Fullview. You can escape anytime you wish."

"Dr. Peachey," one of the technicians said, "how does this *Larry* know about nonspace?"

"Larry is an extension of Betty Jean," Peachey replied. "So he knows everything Betty Jean *wants* him to know—and nothing more. But don't worry, she can endow him with everything she needs of him."

"Did you guys see that!" the botanist exclaimed. "This Larry guy's head just looked like an animal head or something for a split second."

"Yes, I saw that too," said one of the technicians.

"This is to be expected," said Dr. Peachey. "We all have to realize that, like us, our brain mates have image access that goes all the way back to the dinosaur brain. So Larry has access to all three layers of Betty Jean's brain. For some reason, Betty Jean's dinosaur brain wanted to get a look at Larry as a dinosaur, but Betty Jean's human brain quickly brought Larry back to his human brain and body."

"Remember, the dinosaur brain provides us with our sexual and aggressiveness drives," Vandervoort interjected. "Betty Jean may have unconsciously done a quick scan of the male dinosaur aggressiveness and sexuality that drives Larry. This is how her brain evaluates the potential power of Larry as a 'knight in shining armor.'"

"Yes, this shifting from one layer of the brain of Betty Jean's animus to another is completely normal," Peachey continued. "When we are looking at potential mates, we unconsciously imagine their aggressiveness and sexuality before anything else, and we all do it all the time. We're just not aware of it. Because Betty Jean is in Fullview, we actually saw her do the unconscious scan of Larry as a dinosaur. It is a sign of Betty's Jean current mental health, that she was able to so quickly take a look and then put it back into its ancient *cage*. When Betty Jean told Larry that she thought something was watching her, it was this unconsciously produced Larry-as-a-dinosaur emerging in her own brain, and it showed itself as a harmless part of Larry's body. But there are other dinosaurs in Betty Jean's brain that might not so easily be controlled. They are also part of her reptilian brain, and these are the ones she needs to worry about."

"Do you hear that screeching sound, Edythe?" said Professor Vandervoort. "It's getting louder."

"Look, there it is!" said Dr. Peachey. "It's another one of Betty Jean's dinosaurs!"

"Good god!" shouted the botanist. "It's huge, and it's racing right for Betty Jean. I think it's a meat-eater, a . . . a *Saurophaganax*."

"I think you're right," agreed Professor Vandervoort. "The *Saurophaganax* was an ancestor of T. rex. The *Saurophaganax* was 'king of the reptilian meat-eaters.' They lived a hundred million years before T. rex and may have been the meanest of all the big meat-eaters."

"Hey," interrupted one of the technicians. "That saurophag, or whatever you call it, is changing into something else."

"Holy cow," said the botanist. "Now the damn thing looks like a *Carcharodontosaurus*. No, wait, now it's looking like a *Giganotosaurus*. What the hell is going on here, Professor Vandervoort? Why does it keep changing into something else?"

"We are seeing, I think, the *evolution* of the meat-eaters," suggested Dr. Peachey. "That sequence of changes we just saw—*Saurophaganax*, then the *Carcharodontosaurus*, then the *Giganotosaurus*—follows the evolution of the meat-eaters. The whole sequence of the evolution of the reptiles is recorded in Betty Jean's brain, and it's playing out in nonspace."

"The *Giganotosaurus* was the largest of the more recent meat-eaters that lived perhaps one hundred million years ago during the Cretaceous period, which came after the Jurassic period," explained Professor Vandervoort.

"Just look at that thing," said one of the technicians. "Its head alone is the size of a Volkswagen Beetle, and the teeth must be nearly a foot long."

"Why isn't it evolving anymore?" asked the botanist. "Why isn't it changing into the next stage of evolution, the T. rex?"

"It's at the height of dinosaur meanness," said Professor Vandervoort. "After *Giganotosaurus*, the terror of the meat-eaters began to decline. BJ's brain can't produce anything worse than the *Giganotosaurus*."

"But why is she stopping with the awful horror of this creature?" asked one of the female technicians. "Why doesn't she let her brain move on to something tamer?"

"Betty Jean is fascinated by it," replied Dr. Peachey.

Yes, Dr. Peachey was right. I was transfixed by this huge nightmare of a creature. It was terrifying me and, at the same time, somehow beautiful in its power and command. It was lean and well-muscled. As it bore down on me, I could feel its power infusing my body. I felt attracted to it. I wanted it to devour me. *What happened to Larry? I wondered. Why isn't he here to help me?*

"What is she doing?" shouted the botanist. "What has happened to her clothes? She is standing there absolutely naked, holding her arms out to the beast!"

"It's Thanatos," said Professor Vandervoort. "The death instinct has taken over the ancient part of BJ's brain. She knows she can't escape the jaws of *Giganotosaurus*, and she is ready to be eaten, eaten alive!"

"But why is she naked?" went on the botanist. "And why does she seem to *want* to be eaten?"

"I think I had better explain this," said Dr. Peachey. "Betty Jean is naked in the nonspace of Fullview because she has returned to the animal form of her human body. When the animal in us realizes that struggle is useless, it instinctively *welcomes* the nirvana or return to the world of peaceful nonlife from which it came. She is intensely happy with the *feeling* that she's going to be eaten alive."

The huge, hot mouth of the *Giganotosaurus* wrapped around my body. I could feel its teeth sinking into my thighs. It felt warm and pleasant, almost sexual. My head was hanging from the side of its mouth. Inside the giant wet mouth, I could feel its tongue pulling me apart. Saliva and warm blood was everywhere on me and in me. *This,* I thought, *is the most rapturous moment I have ever experienced.* THIS IS BETTER THAN A RIDE ON THE PEACHEY TIME SLED! *I am dying, and I wish it could go on forever!*

"Look at her head hanging out of that thing's mouth," said the botanist. "She seems to be swooning in ecstasy as those huge teeth are tearing her body apart."

"But look at her over there in her Fullview workstation," said one of the technicians. "She is absolutely still, just staring, not moving a muscle."

"Ohhh, the joy, the joy of being eaten!" moaned Betty Jean from her workstation.

All activity in Fullview stopped. The room was dead silent.

"She's *gone!*" suddenly exclaimed the botanist. "She's been eaten!"

The *Giganotosaurus* began licking its chops and looking around for more snacks. Then the weirdest thing happened.

"Professor Vandervoort," Peachey said, tugging on his sleeve. "Look, Larry's back."

"Aha," said Vandervoort in an obviously happy tone. "BJ has survived."

"But where *is* she?" asked one of the female technicians. "I just saw her get eaten alive!"

"You saw her get eaten alive *in her own mind*," said Vandervoort. "Now, you are seeing the return of her mind in the form of her brain mate, Larry. The human part of her brain wants to go on living. She'll use Larry to come back to into Fullview."

"That's right," said Dr. Peachey. "While the animal part of Betty Jean enjoyed being slaughtered and ripped apart, the life instinct of the human part of her is fighting against the animal instinct. This is why brain mates evolved in our minds. Brain mates evolved in the caring mammal part of our brain to save us from our own dinosaurs. They are the mammalian beginnings of *romance*!"

"Hey, look, Larry's turning into a reptile again," said Headley. "Only this fucking time he's not flickering back to his human form. His complete body is turning into a . . . a . . . what did you call it, Professor Vandervoort?"

"Larry has turned into a *Giganotosaurus*, Jay," replied Vandervoort. "But Larry is a male of the species. He probably outweighs the one that ate BJ by a couple of tons."

"Remember, everyone, Larry is Betty Jean's animus," Peachey said. "She is striking back at her female *Giganotosaurus* with the only weapon she has—her male *Giganotosaurus*. This is the real test of whether Betty Jean can come through this without severe mental consequences. If Larry defeats BJ's female *Giganotosaurus*, she'll have defeated it too. If Larry loses, she will have to stay in her present state—eaten alive. Her human mind will have no basic instincts to support it. She'll be lost, and she'll be without Larry, her protective animus."

"Is there any way we can help Larry defeat that goddamn fucking female monstrosity?" asked Headley, for the first time supporting the presence of Larry.

"No!" exclaimed Vandervoort. "The only thing we could do is pull her out of her Fullview computer station and remove the electrodes. But her brain would retain the experience of being devoured alive. The only thing that can save her mind is if she herself works through Larry to defeat the ancient part of her that ate her. You know, the dinosaurs inside all of us are trying to devour us every minute of our lives. And we spend our whole lives trying to defeat them. Betty Jean has a chance, perhaps the first chance in human history, to actually defeat her dinosaur past once and for all."

Larry, in the form of the *Giganotosaurus* within me, attacked the dinosaur that had eaten me. It was a horrible tangle of blood, flesh, and bones. The battle raged on for what seemed like an eternity. Larry's huge *Giganotosaurus* mouth finally crushed the smaller head of his prey.

"Larry's eating the beast that ate Betty Jean," said Peachey.

I felt a new warmth come over my body. It felt like the sun was shining out from my heart. I could see Larry standing beside the body of the dinosaur he had defeated.

"And look!" shouted Headley. "Now, Larry is reverting to his human body."

"Betty Jean's back, everyone!" shouted one of the female technicians. "But are my eyes deceiving me? She looks different. She's, she's . . . even prettier than she was before! How can that be?"

"Well, take a look the fuck over at the Fullview workstation, my little shithead," said an excited, emotionally out-of-control Headley. "*That* Betty Jean has changed a little for the better too. Now, goddamn it, I can see how the Betty Jean in the nonspace can change her looks. It's because *that* Betty Jean is an electronic image coming from her own mind. But how the goddamn, fucking hell can the Betty Jean sitting over there in the Fullview workstation change?"

"She's had a complete mental change," Peachey said while peering intently at the *new* Betty Jean reclining quietly in the Fullview. "Her mental change is so profound it has changed her whole muscle tone and color. And there is an intense light in her eyes. She has entered a completely new state of mind that is not unlike what happens to a gifted impressionist who mimics celebrities. Their faces contort, their eyes take on a new personality—they enter a new reality."

"She is now a human being that has complete dominance of her most basic animal instincts," added Vandervoort. "The old BJ is *dead* forever. To be born again, something has to die. From now on, she will have no fear of death, no desire to have what others have, and no hungers that she can't control completely. In many ways, she reminds us of all the great religious leaders down through the ages. When she allowed herself to be eaten alive and then allowed herself to be saved by her own mysterious brain mate, she shed the deep animal shadow that surrounds all of us. That is the only chance we humans have to transcend into a higher reality. And we have witnessed a true conversion of a human being to a *fuller* human being."

"Professor," interrupted Headley, "I hope you're not saying that she has become some sort of Jesus Christ. Are you?"

"Yes, of some sort," replied Vandervoort. "But don't take that thought too literally, Jay. It is possible, you know, that nonspace has the capacity of building many things entirely new. We can only watch and see what this *new* BJ does now."

CHAPTER 10

The Journey 2: Caress Me—
The New World Order

(2014)

As I sat in my Fullview workstation, I contemplated the incredible transformation I had just gone through in nonspace. Since I had met my death at the hands of my dinosaur brain and then defeated it, my brain and my thoughts were gliding through information so fast I felt like a high-speed computer. I could see now that as long as my dinosaur brain had been in control, my whole reality had been a weird little boxed reality. It had been a compressed life of ritualistic thinking. The fears and angers of the dinosaur brain compress your existence into a shoebox. You don't really see. You don't really hear. You don't think. You don't imagine. You don't really live. That's the big realization! No one with an unpredictable dinosaur brain is actually *living*—no one, not even the pope, no one! Maybe Professor Vandervoort is right. With my dinosaur brain completely under control and only doing my bidding, maybe I *am* some sort of Jesus Christ. Is this what it's like to think and do the kinds of things he did? How did *he* get rid of his dinosaur brain? This *new* me was about to go back into nonspace so that Peachey and Vandervoort could get the next level of computer algorithms for their Living Windows 2016. I wondered what Betty Jean "Christ" would find.

Tomorrow, I was to ask my Fullview workstation to initiate Journey 2, the journey into my ancient *mammal* brain—the world of ancient horses, cats, dogs, and apes. I felt warm thinking about it. What would I find in my early mammal body? Would it be a horse or would it be an ape of some sort? My dinosaurs ate me and tore me apart because that is what they do. I enjoyed it. God, I enjoyed it! But what new *carnage* came into being with the horse and the ape? What would my horse or ape do to me? Whatever they do to me, like the dinosaurs before them, they had to do it to survive! Suddenly, it struck me!

"Why are mammals, mammals?" I muttered aloud. "What allowed them to survive?"

"They survived because they suckled," Peachey announced as she slid up beside my workstation and gently patted the top of my head. "That's why they're called *mammals!*"

"Ah, I didn't know anyone was listening," I responded, somewhat startled. "I was thinking about my experiences with my dinosaurs and thinking about what tomorrow might be like."

"Yes, *tomorrow*," Peachey replied with a serious look.

"Is something wrong about tomorrow?" I asked hesitantly.

"No, nothing is *wrong* about tomorrow," Peachey replied.

"Let me be blunt, BJ," she went on. "You have been more than Professor Vandervoort and I—and Headley, believe it or not—could ever have hoped for. Your powerful mind and beautiful brain have gotten us through to the real world beneath everyday human experience."

I could tell by the tone of Peachey's voice that something *was* wrong with tomorrow.

"But, honey, we have a favor to ask," she continued. "It's about Pauline."

"Pauline wants to take my place tomorrow," I quickly interrupted. "Is that it?"

"In a nutshell, yes," Peachey replied, very obviously relieved that I had anticipated what was coming.

"She was of course very moved by what she saw deep inside your brain," Peachey went on. "Pauline wants with all her heart to go there. She is not concerned about the confusion that her inner body originally experienced in her early days with Fullview, and she is willing to go no matter what happens."

I was surprised that I felt as if I could see deeply into Pauline's wish to personally experience some of what I had experienced. It was a good feeling to think that I had made it possible for someone else to be as happy as I had become. Perhaps this is one of the ways my brain mate helped me to this higher level of reality! Without my dinosaurs eating at me, they can no longer come forward and make me angry, jealous, and vicious.

"What can I do to help you get Pauline ready?" I quickly replied. "I am truly happy that Pauline will be taking my place for Journey 2 into nonspace."

"Pauline is completely ready," Peachey replied as she leaned back against my workstation, giving me that familiar smile of connection between us. "Pauline has been prepared for over a year, but after seeing what your experience was like, she decided that the time to go is *now*. She definitely wants you to be here at the workstation in the morning to 'see her off.' And she wants you in the audience at Fullview with the rest of us. Pauline said she would be reassured if you were there in case she got into some kind of trouble."

"How sweet of Pauline to want me there," I responded. "She must know I wouldn't miss the journey into her brain for anything!"

Peachey, Professor Vandervoort, and I had dinner with Pauline and her mother that night. We celebrated what was about to happen. Pauline glowed at the prospects of the next day, and I glowed *for* her. Then, WHAM, it was the next morning. I met Pauline and Peachey at the Fullview workstation. Over at the Fullview monitor, all the people who were there during my Journey 1 were assembled. Headley, the other technicians, and the scientists were all there, but I noticed a few unfamiliar faces too.

"Who are those new people at Fullview?" I asked Peachey. "Who's the cute gal talking to Headley?"

I didn't know why the girl talking to Headley had so powerfully caught my attention.

"She's the old girl friend who Headley suggested could read the lips of Pauline's homunculus," replied Peachey. "Didn't I mention her in my lecture at Nano-soft's Redmond conference?"

"Oh, yes, yes, I remember you mentioning her," I replied. "Her name was Kimberly, the same as my mother's—that's how I remember it."

"Headley suggested we bring her back for this," Peachey went on. "Professor Vandervoort agreed that it was a good idea in case Pauline's homunculus reappeared only 'in the raw,' so to speak. Kimberly's lip-reading might be helpful in allowing us to talk to Pauline in case her homunculus is all we get."

"I see," I replied. " I doubt, though, that Pauline's homunculus could possibly appear only in the raw homunculus form. In Journey 1, my inner body appeared as my beautiful, deeply desired body. And it had apparently done this through the imagination software of my own cerebral cortex. Wouldn't Pauline's powers of imagination do the same?"

"Possibly not," replied Peachey. "Remember, Pauline has been blind from a very early age. She might not be able to summon an ideal grown-up female body for herself. Her imagination for her own body may be *stuck* in that early homunculus version of her inner body's development. Although, BJ, she did see the example of your beautiful inner body in Fullview. We'll have to wait and see if that one time with your luscious beauty made a deep-enough impression on her."

Pauline momentarily smiled to herself at overhearing Peachey's comments to me and then immediately trained her miniature video camera on Dr. Peachey.

"May I initiate Journey 2, Dr. Peachey?" said Pauline.

"Go ahead, my dear," responded Peachey. "Enjoy yourself. We'll be watching your every move."

Pauline trained her video camera in a quick glance over at the crowd assembled around Fullview. She closed the glass door to her workstation and initiated Journey 2. In an instant, a shadowy image of Pauline popped up in the nonspace of the giant Fullview monitor.

"Headley may have been right," exclaimed Professor Vandervoort. "Pauline may only be able to appear in the raw form of her homunculus. We might, indeed, be needing the services of our lip-reader."

The raw, dark-skinned homunculus of Pauline's inner body assembled slowly before us in Fullview.

"Holy, good god!" shouted a psychologist who was standing near Professor Vandervoort. "So that's what my patients really look like."

"That's right, Dr. Koziol," said an obviously humored Peachey. "You're seeing what we all look like inside. It's the homunculus who lives in all of us, who does everything for us, who comes to visit you in your clinic, but we never realize we are talking to."

The room fell silent as the visiting technicians and scientists watched Pauline's homunculus begin to move about. To us, it was simply Pauline. To them, it must have been a bizarre, frightening sight, just as it was for us at first.

"Dr. Vandervoort!" one the technicians down at the far end of Fullview suddenly called out. "Look what is happening to Pauline's huge hands. They seem to be shrinking and smoothing out."

"Yes, Dr. Vandervoort," I added. "Her huge homunculus lips are shrinking too!"

As we all stood silently in awe, Pauline's sensuous, distorted homunculus body slowly began to take the form of an adult female body.

As Pauline's human body continued to form, a whole new world appeared in Fullview that was quite a little less hectic than were my first moments in the land of the dinosaurs. The world recorded at this more recent level in Pauline's brain had become a different place from what the deeper levels of my brain had produced. The foliage wasn't so huge; most of it was in the greens that we are familiar with today. It was a lush forest. There was a light mist gently flowing in the air.

"I can hear water running" were Pauline's first words in Fullview. "And I can hear birds singing."

I was surprised that the first things she told us about were things she was hearing.

"Hey, has anybody noticed?" yelled out Headley. "Pauline's glass frames and her miniature video camera are gone!"

"Yes, they are gone," Vandervoort said loudly, addressing everyone. "Even though she is blind, Pauline doesn't need her video camera now that she is in the nonspace of Fullview. She is now looking inward to her own ancient world of permanent memories. She has entered her All Time. In the nonspace of the ancient levels of her own brain, she can see as well as you and I can, maybe better."

The light mist enveloped Pauline's body. The feeling of the mist surrounding her made her look down at her body.

"Oh my god!" exclaimed a shocked Pauline. "I am so embarrassed to see that I am completely naked. I don't mean to be naked in front of all of you!"

"It's all right, Pauline," said Professor Vandervoort. "We're all adults here, I think."

"Why am I completely naked?" asked Pauline. "Betty Jean wasn't naked when she first appeared in Fullview. I distinctly remember she was dressed in an army uniform. And my body looks different to me. I must admit, I am shocked at how good I look. My breasts are absolutely beautiful! My nipples are larger than before. How totally embarrassing. Everything on down to my toes is as beautiful as my breasts. Good god, listen to me; I am so thrilled, I am totally thinking outloud!"

"Good god, is right!" exclaimed Headley. "That is the most beautiful female body that could ever be. Jesus, look at those knockers—holy, holy Jesus. *Wow*!"

"Here we go again with Headley," snapped Peachey. "Headley, I can appreciate what you're seeing more than you might guess, but can we please keep this on a *civilized* level?"

"Yes, we'd all better get into a very *civilized* mood," interrupted Vandervoort. "We're likely to see things a lot more interesting than her body come pouring out of Pauline's ancient mammal brain. And some of it might get more *romantic* than having a chat with a brain mate about dinosaurs. Remember, this ancient mammal brain is the part of the brain that gives us humans our unique brand of sexuality. It's the part of the brain that first got nipples to deal with."

"What exactly do you mean by 'romantic,' as you put it?" asked one of the smirking female technicians. "You're not suggesting that the mammal parts of the brain—nipples, if you wish—are connected to romance, are you?"

"In a way I am," Vandervoort replied with a smile. "But remember the ultimate goal of what we are trying to do here. Pauline has to work through all of the deepest memories stored in her mammal brain so that we can record those image algorithms for Nano-soft's Living Windows. One way or another, everything she does in Fullview will involve her mammalian or 'mammary' existence. This will have to involve play, nursing, humor, and deep caring. Some of the things she will do will involve a pure extended sexuality, extended to include both dinosaur and mammalian sexuality. You can call this protracted sexuality, *romantic*, but I don't think I'd characterize it that way. It is better to say that it is the basis for all caring. The reason that people are strongly attracted to romantic sex is because it appeals to both of the original dinosaur and mammal parts of the brain inside them. Through deep romantic situations and books, people can experience the time when they were driven by the image algorithms of both their dinosaur sexuality and those of their mammalian caring—the time when they began to care about the inner lives of others. We believe understanding the inner lives of others is a critical link to how Living Windows will work."

"From our studies of BJ's permanent record of her ancient memories in Fullview, the combination of the dinosaur and mammalian layers of the brain appears to be a different, refreshing sort of a world we enter into," added Peachey. "Romance takes us back to a pure time, a good time. It was a beautiful time. Any real beauty is, after all, a part of nature just *as nature is*. Animals see, feel, and hear a different kind of beauty from what we humans do. They just have no way to express it. When we humans have total, unbridled romantic sex, we step back into that beauty, back completely into nature."

"That's right," continued Vandervoort. "I suppose that's why society will never be able to stop becoming transfixed while reading romance novels or watching soap operas or even, I guess, for some, getting addicted to pornography. And we really shouldn't *want* to stop doing these things. These fixations are simply an ancient part of the brain bringing forward its romantic version of sexuality into our modern world. So our love of romance and reading romance novels is proof that we are largely driven by the ancient mammalian needs within our brains. In terms of what you have been seeing in Fullview, you can think of all of this romantic sexuality as a *mental* form of *zoos* that our own brain can create for its own amusement. The modern human brain is sort of a romance-starved puppeteer of the animal puppets that live within it. Its mental zoos are better than physical zoos. They have only *mental* walls around them and can be visited anytime and to any degree we wish. Hopefully, for the Living Windows project, BJ and Pauline will have shown us *all* of these mental images in these zoos so we can harness them and make them work in Living Windows."

"So are you're telling us that Pauline's going to have to go to the max with romantic sex just like BJ did with being *eaten* alive when she was in her dinosaur world?" asked a slightly incredulous, slightly hopeful Headley.

"That's right," answered Vandervoort. "It's the only way we can record *all* of the image algorithms in the mammal brain. We have to go the max, as you put it, with Pauline just as we did with BJ."

"But, Jesus fu—whoops, sorry, Dr. Peachey, I don't want to turn you on. Ha ha," Headley exclaimed. "But *my gosh's sake*—I hope that's civilized enough for you—BJ *enjoyed* being eaten alive in the dinosaur world," said Headley. "What's Pauline going to do with romantic sex, *enjoy* getting loved and caressed to death right here in front of us?"

"Goddamn it, Professor Vandervoort!" screamed a highly agitated Peachey. "Would you please get that half-wit out of here? We're headed into some stuff that's going to be pretty difficult for us all to deal with. We can't risk the success of these sessions with that kind of bizarre behavior going on. Headley's getting way out of control."

"No, Headley, I don't think Pauline's going to get caressed to death right in front of us, as you put it," Vandervoort calmly answered while momentarily ignoring Peachey. "I know you'd probably like to see that, but her confrontation with pure romantic, caring sex might not be what any us would expect it to be. We'll just have to wait and see what happens."

"Professor Vandervoort," said Peachey, again attempting to get a response about Headley's behavior. "Are you going to get rid of Headley or not?"

"Headley, why *don't* you just slip around to the other side of Fullview and get your *full view* of Pauline's *attributes* from there?" said Vandervoort, putting his strong hand on the top of Headley's head. "You can also keep a closer eye on Pauline sitting in her Fullview workstation from there. I'll call you back over if we need you."

"It's your show, Professor," said a lightly smirking Headley as he jumped over a row of chairs and began to disappear behind the low cradle of beams that held Fullview in place. "I'll be watching Pauline get caressed to death, Dr. Peachey. Hope you enjoy it as much as I will!"

As Pauline sat in my workstation, she watched the workstation monitor through her miniature video camera as her face and body slowly changed in the larger Fullview monitor. I knew It was strange for her to be able to watch from the workstation and to watch from inside the larger monitor at the same time as I had done. I also knew she could see Headley over behind Fullview, peering back and forth at her sitting in the workstation and then at her walking completely naked in the larger Fullview.

"Look, Pauline is walking along the shore of a lake," said one of the geologists. "That shoreline must be a mile wide, and look, you can see mountains off in the distance on the other side of the lake. Where is she, anybody know?"

As we watched, the shoreline grew wider and wider until the lake completely disappeared. It became an arid landscape. Then the lake was suddenly back and washed up against Pauline's feet. Then the shoreline quickly began to widen again.

"Yes, as a matter of fact, I think I know exactly where she is," replied Professor Vandervoort. "Since we are getting images from the late mammalian development of Pauline's brain, and since we are seeing huge lakes rising and disappearing, she must be reliving the ancient climate changes of the East African Rift area. The rapid climate changes were severe, and only the most cunning and adaptive groups survived. These climatic hardships slowly shaped the brain of *Homo erectus* by about two million years ago. It occurred over countless ten-thousand-year cycles and became permanently recorded in all our brains. We are seeing the repeated do-or-die 'crunch times' that made the mammalian brain slowly take its next steps toward more brain and better brain."

"I'm in East Africa two million years ago?" asked Pauline in astonishment.

"I think that's exactly where you are," replied Vandervoort.

"I'm completely *naked* on the shore of a lake in East Africa two million years ago?" asked Pauline, repeating her situation and upping her concern.

"Well, you're not completely naked anymore," I interjected with a sigh. "Have you looked at your legs lately?"

Pauline looked down at her legs and was obviously shocked to see that they had become more *masculine* and lightly covered with hair!

"What has happened to my legs, Dr. Vandervoort?" Pauline asked. "I don't remember anything like this happening to BJ when she went back to the land of the dinosaurs in Journey 1."

"Don't worry about your *legs*, Pauline," interrupted Headley before Vandervoort could speak. "Your entire body is lightly covered in the same thin, long hair!"

"Your brain is turning your body into what it saw and recorded as the female *Homo erectus* of that time," replied Vandervoort. "Pauline, you are now two million years old!"

"Yes, Pauline, you *are* two million years old," added Peachey. "But remember, we are only retrieving the *hairy* you from the permanent record in your brain. Look over at your workstation. The *you* as you know yourself is still here with us. Every one of us around Fullview has images of *Homo erectus* in the mammalian part of our brains. That part of each of us is two million years old."

"Pauline," said Vandervoort. "The reason Betty Jean appeared as a beautiful modern female when she went to the land of the dinosaurs is simply because there were no ancient humans like *Homo erectus* during that time. Remember, the holographic image-gathering electrode was recording from her *reptilian* brain. There are no images of ancient humans in the reptilian part of our brain, so the only way BJ could appear in Fullview was as her modern brain could visualize her."

"That hair is kind of sexy, Pauline," yelled Headley from the other side of Fullview. "But isn't she kind of tall for one of those ancient, two-million-year-old hairy humans, Dr. Vandervoort? I thought they were only supposed to be about three or four feet tall."

"Good observation, Mr. Headley," said a paleoanthropologist who had come in to witness Pauline's journey into her mammalian brain. "But no, *Homo erectus* might well have been as tall as Pauline. One skeleton of a twelve-year-old male *Homo erectus* we found was five feet, three inches tall. If he had grown to manhood, he may well have been six feet tall. You may be thinking of *Australopithecus*, a pre-human who came along before early humans and was much shorter."

"Dr. Vandervoort," again yelled Headley, "Pauline's skin is dark like it was when we first saw her homunculus. Do you remember that? You said it was because her homunculus was formed in Africa long ago."

"Yes, I remember, Headley," replied Vandervoort. "Apparently the *Homo erectus* group recorded in Pauline's brain had not yet lost its dark skin color by two million years ago. Two million years ago,

Homo erectus had not yet migrated to Europe and the Middle East, where new conditions and mutations made their skin color begin to slowly change."

Pauline entered a trail that was leading away from the lake shoreline and into the surrounding lush forest.

"I am hearing a rhythmic clattering in the forest," said Pauline. "I am intrigued by the sound. I am going to go there. I must go there."

Pauline followed the path for what seemed several hundred yards, where the forest then began to open up. Suddenly, the clattering stopped. It was eerily silent. As Pauline continued to move cautiously forward, a large pit with smoldering embers appeared in the middle of the clearing.

"Somebody has had a fire in that pit," said one of the botanists. "I thought Pauline was back two million years. Did early humans learn to use fire by two million years ago?"

"Fire actually fits in pretty well two million years ago," replied the paleoanthropologist. "*Homo erectus* had developed crude stone tools by then, and we know they had a lot of meat in their diet. They had probably been cooking their meat and plants for at least a thousand or two generations before this got recorded in Pauline's brain. Remember a thousand generations for these *Homo erectus* guys was less than twenty thousand years, a small amount of time in an overall two-million-year period."

"And that thousand or two generations of meat eating and cooked-plant eating could have easily added an inch or two to their height. It also could have done wonders for their brain development," added Vandervoort.

"Yes, and don't forget, Dr. Vandervoort, that those thousand or two generations occurred during glaciations and interglacial periods," commented the paleoanthropologist. "These huge climatic variations would have caused equally great plant and animal changes that meant only the stronger and smarter erectus groups would have had the wherewithal to survive. Over a period of a thousand generations, certain groups of *Homo erectus* could have made great strides not only in their size, strength, and brain power but in the way they lived. In some local niches, they might have become quite comfortable and quite socially interesting during the interglacial periods."

"There they are," announced Peachey. "Will you look at that? We are seeing real *Homo erectus* women for the first time. And, Dr. Vandervoort, you were right back when we first saw Pauline's homunculus about why it was dark skinned. These erectus women are dark skinned, just like Pauline's homunculus."

"This is amazing!" said the paleoanthropologist. "The erectus women look pretty much like we had thought they might look. Everyone, please notice that they look nothing like the half-human, half-ape portrayed in *Planet of the Apes*. We knew erectus wouldn't look apish. Unlike those half-and-half, apish-looking *citizens* of *Planet of the Apes*, erectus evolved from another line of pre-humans that were quite different from the apes we are familiar with today. These erectus women are not apes at all. They are early *people*."

Just beyond the fire pit, a dozen or so erectus women sat, staring at Pauline.

"They have never seen a creature like Pauline!" yelled Headley.

"And neither have you, huh, Headley?" Peachey yelled back partly in humor, partly in sarcasm.

"Why are they just staring at me?" asked Pauline. "They aren't moving a muscle! This is creepy, very creepy."

"Pauline, you just popped up out of nowhere to them," replied Peachey. "They might think you're kind of *creepy* too."

"Not to mention, again, the fact that they have never seen a *female* quite like you!" added Headley.

"Yes," said Vandervoort, "you are taller, more slender, and to them, your modern human face may seem to be a very strangely distorted face. They probably don't know what to do, and they are probably afraid of you."

"And you look good, like a *mammary* animal should," yelled Headley with a chuckle. "Right, Dr. Peachey?"

Slowly, the erectus women began to drop the large stones they were holding and, keeping their eyes on Pauline, began to stand.

"Did you take close notice of the stones the women are dropping?" said the paleoanthropologist. "They are what we call Oldowan stone tools. These tools were named after the Olduvai Gorge in East Africa, and it appears we have been right that erectus women and not the men were the first major toolmakers. Oldowan tools were very primitive and used mainly for meat and plant cutting, mostly things the women did. That rhythmic clattering we heard was the women hitting the stones together to make more tools."

"Toolmaking was so important to survival for erectus that the mental and manual skills behind its rhythms shaped their thinking, their utterances, and their body movements," added Vandervoort. "Those who could quickly learn to hear the right sounds when rocks were struck together could make the best tools, the hardest and sharpest tools. And those tools meant survival. They developed an excellent feel for the fracture mechanics and internal geometry of fine-grained lava, quartzite, obsidian, flint, and limestone. The nuances of the highly repetitive toolmaking sound rhythms came to bond them together and led to their chanting, their calls, and what could only be called the earliest beginnings of *music*. Stones pounded upon one another by the erectus people were the first rough musical instruments. This early *music* came a million and half years before language, but I think it began to provide the basis for the complexity and rhythms of language. By mixing with utterances and vocal calls, primitive rhythms set early humans on the course toward the evolution of the languages we speak today."

One of the larger erectus women, perhaps five feet tall, had been carefully studying Pauline's strange, beautiful body. Suddenly, she let out a loud screaming call. It wasn't exactly an animal call; it was a rhythmic and repetitive "whoa-boo-boo, whoa-boo-boo, whoa-boo-boo!"

"What should I do, Dr. Vandervoort?" said an obviously frightened Pauline.

"Stand your ground, Pauline," replied Peachey before Vandervoort could speak. "If you appear frightened, they may attack."

"I am very frightened," replied Pauline. "I can't control my fear."

"Are you seeing what I am seeing from over here?" yelled Headley.

"My god!" said Peachey. "Pauline is transitioning into the form of her inner body, her homunculus!"

In a matter of seconds, the huge hands, lips, and tongue of Pauline's homunculus replaced her beautiful, outer body. The erectus women, who were beginning to approach Pauline, immediately recoiled in astonishment and fear.

"I see that my homunculus has come back!" said an excited Pauline, sitting over in her Fullview workstation. "Why has it come back? Is it confused again, Professor Vandervoort?"

"You are afraid, Pauline, very afraid," replied Vandervoort. "You unconsciously know the erectus women are threateningly curious about your beautiful face and body, and your homunculus has stepped forward to hide your beauty from them."

Now, even more frightened by the sudden appearance of Pauline's strange homunculus, *all* the erectus women began to loudly repeat the rhythmic predator chant "whoa-boo-boo, whoa-boo-boo, whoa-boo-boo!"

"What are they doing with that chanting?" yelled Headley from across the room.

"It seems to be some sort of advanced call that is telling everyone in the erectus community that there is a *predator* confronting them," replied the paleoanthropologist. "I predict we'll be seeing some erectus men showing up before too long."

Pauline's homunculus was looking around at the ground as if it had an uncontrollable urge to sit down so that the erectus women would see that she meant no harm.

"Pauline's homunculus is sitting down," said Peachey. "Her inner body is reacting to the predator chanting of the erectus women. Look, she is reaching her huge hands out to them."

"Pauline's homunculus is facing me now," yelled Headley. "From over here, I can see her huge lips moving."

Headley waved his arms to Kimberly, his deaf lip-reader girlfriend, to get over to his side of Fullview pronto! Kimberly quickly appeared next to Headley. Kimberly knew why she was there and immediately began to study what the huge lips of Pauline's homunculus might be saying.

The huge hands of Pauline's homunculus continued to motion to the erectus women to bring the stones they had been shaping over to her. Headley's Kimberly continued to study the movements of Pauline's huge lips.

"Headley, can Kimberly decipher what Pauline's lips are saying to the erectus women?" yelled Vandervoort.

Headley tapped Kimberly on the shoulder so she would turn to him to read his lips.

"Are you getting anything, Kimberly?" Headley asked.

"Yes, she is repeating the same lip movements the erectus women make when they chant their predator call," Kimberly loudly replied so we could all hear. "She is doing their 'whoa-boo-boo, whoa-boo-boo, whoa-boo-boo!' call."

"That homunculus of Pauline's is brilliant!" said Vandervoort. "Pauline's homunculus is trying to calm the women by saying something to the effect that 'I join you in making your predator call—I am one of you.'"

"Can the erectus women hear what Pauline's homunculus is saying?" asked one of the technicians.

"I don't think so," replied Vandervoort. "That predator call the women make is so important to them they must have intuitively realized exactly what Pauline's homunculus was saying by watching her lips. Say 'whoa-boo-boo' and notice what happens to the shape of your lips and how your jaw moves rhythmically up and down. The erectus women recognized those movements in the huge lips and tongue of Pauline's homunculus. They automatically knew her lips and tongue were giving the predator call."

"I agree, Dr. Vandervoort," added Peachey. "The erectus people were probably much better than we are at interpreting the fine details of images of what they saw in Pauline's huge moving lips and tongue. And did you all notice the extreme facial expressions that erectus women use when making the 'whoa-boo-boo' call?"

"Yeah, they really get into it, don't they?" Headley said with a smirk.

"When they communicated with one another with their rhythmic sound patterns, they were purposely sending and *hearing* images, not words—they did not yet have anything like language as we know it," Vandervoort went on.

"I agree too," said the paleoanthropologist. "By lip-synching the 'whoa-boo-boo, whoa-boo-boo, whoa-boo-boo' chant of the erectus women, Pauline's homunculus has cunningly calmed them down. And look, there's the proof. The erectus women are bringing their stones to Pauline's huge hands."

"Hey, everybody," yelled Headley. "Take a look at Pauline sitting over in her workstation. That's quite a mischievous smile she has on her face! She looks pleased that she just pulled a fast one on *Homo erectus*!"

"I wasn't grinning because I pulled a fast one on these guys," said Pauline from her workstation. "I was grinning to myself about how easy it would be for me to show them how to make better stone tools."

"How would you do that, honey?" asked Peachey in friendly disbelief.

"My mother enrolled me in an anthropology class not far from Redmond at Bellevue Community College, where we learned to make all sorts of stone tools," replied Pauline. "Although I couldn't see what I was working with, I can remember the feel of the different types of stone tools that developed over hundreds of thousands of years. I particularly remember the Oldowan tools and the later, more sophisticated Acheulean hand axes. Now that I am in Fullview, and I can actually see the Oldowan tools these erectus women are making, I know I could move them ahead a few hundred thousand years to the stone tools of Acheulean times in a few minutes. Dr. Vandervoort, is it okay if I teach them how to make Acheulean hand axes? It would make life a lot easier for them. And they might learn to *like* me."

"Maybe you'd better not!" yelled Headley from the other side of Fullview. "You don't want to change history by catapulting these people ahead of their time! You teach them how to make Achu-whatever hand axes, and we might all suddenly disappear because you changed the entire pace of history."

Everyone assembled around Fullview was silent.

"Go ahead and teach them whatever you wish, Pauline," replied a laughing Professor Vandervoort. "Remember, everyone, all of this that's going on in Fullview is only going on in the permanent record archived deep in Pauline's brain. Pauline can get married and have children, spend the summer or whatever she wants while she is with *Homo erectus*. It will not affect history or us in any way, although getting married might have an effect on the Pauline sitting over there in her workstation. It might affect her in ways similar to how Betty Jean was affected by being eaten alive and then saved from her dinosaurs. It could be therapeutic, somewhat like visiting a psychologist. After all, that's what psychologists really do; they straighten out the ancient impulses of your mind."

Pauline seemed relieved that the erectus women had calmed down and were timidly approaching her homunculus body with their stones. The erectus women now, in a naïve way, seemed intrigued with Pauline's greatly distorted homunculus.

But just as the erectus women began placing their stones within reach of the huge hands of Pauline's homunculus, the homunculus began to transition back into Pauline's beautiful outer body. Surprised, the erectus women stepped back in renewed astonishment.

"Pauline has her fear under control," said Peachey. "Thank God she's back!"

Pauline continued to sit as she reached for two large hunks of glassy lava the erectus woman had brought forward. She kept her head down while she used one of the stones to break the other one in half. Pauline then began quickly knapping chips off the smaller stone. The erectus women slowly began to approach and gather around her.

"Look at her go!" exclaimed the paleoanthropologist. "She is really good at stone tool knapping!"

"Yes," Peachey added. "Remember, because she was blind when she learned knapping at the community college, she has an excellent feel in her hands and highly tuned ear for the fracture mechanics and percussion sounds involved in stone knapping."

"And remember too that because Pauline is looking down through the layers of her own brain while in Fullview, she has twenty-twenty vision to increase even further her stone-knapping skills," added Vandervoort.

Pauline soon completed a rough hand axe approximately six inches long with sharp, hard edges. She held it out for the woman and motioned to them to take it.

"Interesting!" I commented. "The erectus women seem curious about the hand axe Pauline has made for them, but at the same time, they seem puzzled by its triangular shape."

"Yes, Betty Jean," said Vandervoort. "To them, it's like when we get a new piece of computer technology or a new kind of iPhone and don't know all the things it can do. The erectus women have no idea what this new Acheulean *device* can do or how to use it."

"Look out, Pauline!" yelled Headley. "They are coming up behind you, and they're coming fast!"

Pauline turned to see several large *Homo erectus* men rapidly approaching. They threw the several small dog-sized animals they had caught to the ground as they menacingly moved forward. Pauline quickly wheeled around and stood tall while simultaneously raising the hand axe high above her head.

"Dr. Vandervoort," said Peachey, "those erectus men are easily the size of modern humans. Some of them are taller than Pauline."

"Look, the erectus women are surrounding Pauline and facing outward toward their men," said the paleoanthropologist. "They seem to be trying to protect Pauline."

With the erectus women between them and Pauline, the erectus men suddenly stopped their approach.

Pauline quickly strode over to the carcass of one of the small animals the erectus men had thrown to the ground.

"What is Pauline doing?" asked one of the technicians. "Is she crazy?"

Pauline raised the hand axe high above her head and plunged it deep into the animal. The erectus men were apparently so startled by this swift, decisive maneuver they reflexively stepped back.

"No, she's hardly crazy," said Vandervoort. "Believe it or not, she must be on *autopilot* from what she learned in that community college toolmaking class. They obviously not only taught her how to make the Acheulean hand axe—they did an incredible job of teaching her how to use it. And being blind at the time, it all must have made a powerful, indelible impression on her."

"She has not only taught the erectus women how to make the hand axe, but she has just taught all of them one of the ways they can use it!" I added.

The erectus men and women were silent. They stood, looking at the hand axe buried deep in the animal carcass. After a few moments, one of the larger *Homo erectus* men looked away from the carcass and fixed his eyes on Pauline.

"My, look at the way that six-footer is tilting his head back and forth and up and down, taking in Pauline," commented one of the female technicians. "He's ogling her!"

"He's sure a strange-looking guy," said one of the other female technicians. "But he's kind of cute, kind of weird cute, if you know what I mean."

"That *guy*, as you put it, is two million years old, doesn't know about women's lib, and can't speak," yelled Headley from the other side of Fullview. "I think he's dangerous cute, not weird cute! We'd better keep an eye on him."

Pauline, sitting in her Fullview workstation, strained her head to aim her small video camera back at Headley, where he and Kimberly were sitting. At that same moment, Pauline inside the huge Fullview monitor also peered out at Headley and Kimberly.

"That comment sure got Pauline's attention!" said Peachey. "I think this is the first time we have seen both Pauline in her workstation and in her Fullview nonspace representation looking at the same thing at the same time. Headley's comment must have hit a very deep chord."

The erectus women cautiously approached the animal carcass, looking carefully at how the hand axe had quickly penetrated deep into its body. The tall erectus woman who had first given the predator call looked at the erectus men as she slowly pulled the hand axe from the carcass. With her arm extended and the hand axe flat in her open hand, she momentarily displayed it close in front of Pauline and then continued slowly on around, displaying it to the erectus men.

"She is motioning to the men to take the hand axe," I said.

"Yes," added Vandervoort. "And it seems she is telling them that it is a great gift from Pauline to all of them, and that Pauline is now as one of them."

An older, much shorter erectus man moved forward, taking the hand axe from the woman's hand. He examined the hand axe carefully for a moment and then knelt by the animal carcass and, while looking straight at Pauline, plunged it deep into the carcass.

"Ah," said the paleoanthropologist. "Look at them smile, and listen to that humming-grunting sound. They obviously approve of Pauline's hand axe."

"And they obviously approve of Pauline," I added.

"Yes, Betty Jean, Pauline has skillfully made friends with the erectus people," agreed Peachey so all could hear. "More accurately, everyone, she has made friends with the caring and loving forces of the mammalian part of her own brain. Through the cunning of her inner body's homunculus making the 'whoa-boo-boo' predator call when confronted by the women and then her skill of making the hand axe, the emotional and rational parts of her brain have come into sync."

Professor Vandervoort stood silently, stroking his mustache and goatee.

"Do you realize?" said Vandervoort, looking pensively at Pauline and the erectus people in Fullview.

Vandervoort paused.

"Do we realize what, Professor?" I asked.

"Do you realize that in this whole scene we have just witnessed, Pauline showed extreme cunning with nothing but pictures, nothing but images," Vandervoort replied. "Pauline's homunculus mimicked the erectus predator call by moving its lips for the call. That way, she solved a complicated problem of being accepted, of making friends of potential enemies. Pauline used the hand axe to show the erectus people a new invention that could make all of their lives better. It was all accomplished with nothing but images, the series of images the erectus people could only *see*. During the whole thing, Pauline did not have to say anything!"

"That's what you have been telling us all along, Hans," said Peachey. "The human brain is only *smart*, smart way beyond computers, because it uses images in cunning ways. To think and communicate the highest levels of thought requires only the stringing of pictures together from the huge storehouse of millions of years of memories with new pictures constantly coming in from the senses."

"And, Edythe, we have learned something new here for Living Windows," replied Vandervoort. "It is the final piece of evidence for what we are trying to accomplish for Nano-soft! The erectus people, who have no language, understood Pauline's cunning immediately. They kept right up with what she was thinking using only memory pictures from their nonlanguage brains. Even though the erectus people don't have fully modern human brains, they are able to accomplish advanced thinking with just memory pictures! This is the secret to developing Nano-soft's Living Windows operating system!"

"*What* could be the *secret*, Dr. Vandervoort?" yelled Headley from the other side of Fullview.

"The secret is exactly what Dr. Peachey just revealed," replied Vandervoort as he looked over at Peachey. "Living Windows software can think at the same level the erectus people do by simply mixing together memory pictures from the powerful dinosaur drives with memory pictures from the caring of mammals, including those of the erectus people; memory pictures do all of our thinking!"

"I'm not a computer software expert, Dr. Vandervoort," said the paleoanthropologist. "I don't get it. How on earth would it work? How would it be built?"

"We have to back up a bit to understand this," replied Vandervoort. "First, what we witnessed in Fullview with Betty Jean's journey back to her dinosaurs and now Pauline's journey back to the time of *Homo erectus* has been the construction of the human brain as a *living* computer. The *living* computer is sitting over there in the Fullview workstation. Each step of the way during millions of years whizzing by in Fullview, Headley has been electronically recording *all* of the image sequences which have occurred, the ferociousness of the dinosaurs, Betty Jean's brain mate, the erectus people, Pauline's hand axe, everything! These images and the way they are sequenced are the methods or *algorithms* by which the living brain has solved all of its problems. Thinking and problem solving do not come from words. Thinking comes directly from the special way images or memory pictures are sequenced."

"Dr. Vandervoort, may I interject something here?" asked Peachey.

"Sure, go ahead, Edythe," Vandervoort replied.

"You probably don't remember me telling you years ago, Hans, that I did my master's degree in psychology on Albert Einstein's view of how the human mind works," continued Peachey. "Do you remember that?"

"Yes, I do," replied Vandervoort with a smile.

"Einstein, Shmine-stein," said an obviously annoyed Headley. "Nobody is this room has a clue about understanding Einstein except maybe Dr. Peachey. Can we just pass over him for now?"

"Jay, Einstein is important here," responded Vandervoort. "I am sure Dr. Peachey can tell you that Einstein was very interested in the fact that psychology and the principles of physics are totally inseparable."

"Actually, Headley, late in his life, Einstein left some notes that hardly anyone has read," Peachey quickly responded. "He made a point of saying that his thinking was just like everyone else's. Headley, even you would enjoy reading what he had to say about how the brain works."

"Anyway, Hans, Einstein would have agreed with you about how sequences of images are the key to human thinking," Peachey went on. "He said that all of his own ideas were the result of series of what he called memory pictures. These sequences of memory pictures intuitively flowed in his mind until a special picture or image appeared that tied the series altogether. Einstein said that this special picture tied things together and led to his ideas like relativity theory and $E=MC^2$. So Einstein felt that the brain was kind of a living computer that used intuitive sequences of mental pictures to solve problems."

"Aha!" exclaimed Vandervoort. "So the power of the mental images we have found during the journeys in Fullview supports Einstein's view of how thinking works, and Einstein supports the idea that the living brain does its computing by using strings of memory pictures."

"Hey, hey, hey, Dr. Vandervoort!" yelled Headley from the other side of Fullview. "That sounds great, but I didn't hear Einstein say anything about the dinosaurs and the erectus people and Pauline's inner body. Where do they come into Einstein's *picture*?"

Peachey rolled her eyes at Headley's playfully blatant impudence.

"Actually, Dr. Peachey might answer that better than I can, Headley," replied Vandervoort. "But just let me say I remember that Einstein did comment that his discoveries did *not* start with words. Instead, he said his deepest ideas started with 'psychical entities' in the form of images and that some of these psychical entities were of a deep muscular type."

"Yes, that's right, Dr. Vandervoort," added Peachey. "Muscular symbolism intuitively arising from his own body was very important in Einstein's thinking. These muscular, psychical entities Einstein mentioned are very ancient and basic in the brain—they are the muscles of the homunculus in each of us. They go all the way back to the ancient muscle system in the dinosaur part of our brain, all the way back to the dinosaur memories that Betty Jean brought into Fullview. As a very young man, Einstein began his thinking in a thought experiment by imagining what it would be like to take a ride on a beam of light."

"A thought experiment, everyone, is an experiment that is carried out only in your head," interjected Vandervoort.

"Einstein's imagined experience of riding a beam of light was a blend of dinosaur muscular images and visual images, and the blending was done by his modern human brain," continued Peachey. "Later, he used another thought experiment to show how gravity came about. He imagined how his body would float around in a falling elevator. All of this was going on in the nonspace of Einstein's mind."

"And the only *body* that could have the 'imagined muscular wherewithal' to ride a beam of light inside or float in a falling elevator in thought experiments inside Einstein's head is his inner body, his homunculus," again interjected Vandervoort. "In other words, and this is extremely important, there is no one else or nothing else in Einstein's brain but his homunculus that could have imagined the experience of riding a beam of light or the experience of floating in a falling elevator."

"I am beginning to see the *secret* to how Living Windows would work," said the paleoanthropologist.

"Wow, wow!" exclaimed Headley, "I can just see Einstein's homunculus riding the light beam. So if we hooked up Einstein in the Fullview workstation, we could watch him doing the light beam experiment in Fullview. We could watch him riding a beam of light, and time would be slowing down in his head—and it would also slow down in Fullview because there is no space in Fullview—it's the nonspace of the mind!"

"Kind of like a scientific orgasm for Einstein, huh, Dr. Peachey?" I commented. "Kind of like the Peachey Time Sled."

"Yep, Betty Jean," said Peachey. "It's an old Freudian concept. Science is simply sublimated or repackaged sex, dinosaur sex! I think Headley just took a sled ride himself! Wow, wow, wow!"

Everyone around Fullview broke out in laughter.

"Headley, you have actually hit it right on!" said Vandervoort. "While you are recording the image-sequence algorithms from Pauline, you are also recording how her inner body or homunculus is using those sequences to solve problems. The homunculus knows how to use the images to solve both everyday problems and totally new problems exactly like Einstein's homunculus did. The secret to designing the Living Windows software is to give it the inner body or homunculus algorithms you are recording."

"So when Einstein described how he used memory pictures and muscular sensations to solve deep problems, he was actually describing how his homunculus solved problems in nonspace," commented Peachey.

"Yes," continued Vandervoort. "In the nonspace of Fullview, we discovered how Betty Jean and Pauline have solved problems when they were threatened. In each case, they were transformed into the reality of the deep problem solving that Einstein described was going on in his own brain. Betty Jean's and Pauline's electrodes showed that their homunculus came to life to solve a problem. We have witnessed the *Einstein algorithm*. Don't worry about the word "algorithm," it just means that, although Einstein didn't know it, his brilliance was in his ability to look deep inside to see his own homunculus coming to life to solve tough problems! All great ideas come from looking inward, and the inner body, the homunculus, is the one who looks inward into nonspace. It is only in nonspace that thought experiments can be carried out."

"Holographic connections?" yelled Headley.

"You have been recording holographic images all along, Headley," replied Peachey. "You know that the electrodes in both Betty Jean and Pauline are holographic, image-gathering devices. Only *holographic* image-gathering devices have the capacity to collect the entire two hundred million years of images stored in the brain."

"*The homunculus is the one who looks inward!*" suddenly exclaimed the paleoanthropologist. "Now I get how Living Windows would work. The secret of Living Windows is to give it its own homunculus, right?"

"That's right," replied Vandervoort. "We will have to give Living Windows software its own inner body. That inner body or homunculus will experience the world, encounter the problems given to it, and then search and mix its archives of millions of memory pictures for solutions. That's what Einstein said he did. It's the only thing anyone else can ever do, including all of us sitting here around Fullview."

"Speaking of Fullview," interrupted Peachey. "Look, Pauline seems to be the center of a bit of commotion!"

During our brief discussion of the possible nuts and bolts of Living Windows software design, we had been distracted away from what Pauline and the erectus people were doing in Fullview. Many of the erectus men still stood obviously stupefied at the unfamiliar but sensuous new form of beauty of Pauline's face and slim, tall body.

"What is going on in there?" continued Peachey as everyone intently peered into Fullview. "What is going on with that agitated group of erectus men?"

Headley began signing to his girlfriend, Kimberly, who had been watching the Fullview monitor while everyone had been distracted with the Einstein discussion. Headley signed to her to tell him what was going on.

"Yes, I've been watching while all of you have been talking," yelled Headley's girlfriend, Kimberly. "One of the erectus men has been looking at Pauline quite a bit, *quite a bit*. He has been looking her up and down and sneaking closer and closer to her. He's the big one and seems to have a little following of *henchmen*. Pauline has noticed this going on around her, and I have been watching it begin to freak her out."

"Yeah, it's kind of funny watching that group," said Headley nonchalantly. "They are acting like a little gang of modern-day hoodlums."

"Well, Headley, that's quite an astute observation—coming from *you*," said Peachey. "That's probably exactly what they are. They are the 'out-group' among these erectus people. That guy panting over Pauline appears to be the 'gang' leader. He probably wants to have her as a trophy he can share with his 'henchmen,' as your Kimberly calls them."

"I agree, Edythe, they do appear to be the 'tribe's' out-group," interjected Vandervoort. "They are mean, not-quite-socialized young men who are stuck somewhere between their innate dinosaur aggressiveness and sexuality and the very beginnings of early human culture, and they attack in packs like wolves, tearing apart and feasting on their prey in an uncontrolled frenzy all the while because they have acquired a mammalian brain, ironically doing it simply because they feel they have been left out."

"So these aggressive young men are driven to avenge their unfulfilled sexuality with women by stepping out of society and taking what they want?" inquired the paleoanthropologist.

"Yes," replied Vandervoort. "These aggressive men, being mammals, did not get enough loving, care, and reassuring suckling as infants. They are therefore, for want of a better term, *incomplete mammals*. To such men, the female body, especially the mammary nipples, are a kind of stimulus or 'sign' that evokes unfulfilled and therefore untamed aggressive and sexual urges. The dinosaur aggression and sexuality that you saw attack Betty Jean in Journey 1 was not brought under control by the fledgling, suckling mammalian brains of these angry young erectus men during their babyhoods."

"Hans," Peachey said thoughtfully to Vandervoort, "that fits in beautifully with the new research that reveals the rich sexual neural wiring and large size of the nipples in the female homunculus."

"Precisely," replied Vandervoort. "Nipples are the mammalian thing. Deep in the brain, they represent care and caressing and love and belonging.

"Soooooo, Dr. Vandervoort!" yelled Headley from the other side of Fullview. "You're telling us that little group of perverts is after Pauline's nipples? This I gotta see!"

"If their dinosaurs get loose like mine did, those erectus boys will go after a lot more than her nipples!" I exclaimed. "I've been there, done that."

The older erectus man who had plucked Pauline's hand axe from the dead animal and had waved it around for all to see its power suddenly stepped forward. He firmly placed himself between Pauline and the erectus man who had been *stalking* her.

"Ah, help has arrived," said Peachey. "The old man must be the accepted 'wise man' of the tribe, and he is attempting to act as peacemaker between Pauline and the group of troublemakers."

The old man held up the hand axe. He glanced at Pauline and then at the hand axe. Apparently, he meant it not as a threat to the little band of troublemakers but as a symbol of Pauline's contribution to the tribe.

"Look," said Betty Jean, "the 'hoodlums' are stepping back. They seem to respect the power of the hand axe."

"From over here, we see one of the pervert's henchmen sneaking back into the crowd," yelled Headley and his girlfriend almost in unison. "It looks like he is circling around to the side of the old man."

"Tell the old man one of the hoodlums is sneaking up on him, Pauline!" warned Peachey.

Pauline reached out and touched the old man on the shoulder. As he turned to look at Pauline, the henchman jumped out of the crowd and quickly snatched the hand axe from the old man's raised hand. The henchman threw the hand axe to the feet of the large erectus man who had stalked Pauline.

"Uh, oh!" exclaimed Peachey. "The *stalker* is picking up the hand axe, and he doesn't look happy with the old man's attempt to make peace."

The erectus man who had been stalking Pauline held the hand axe for a moment. He ran his fingers lightly over the razor-sharp edges Pauline had so skillfully knapped.

"Look at him!" said the paleoanthropologist. "Look at that devious smile he is giving Pauline as he runs fingers over those sharp edges."

The stalker grunted loudly, glanced at his henchmen, and with his free hand, waved them toward the old man. The henchmen took the old man to the ground, spreading his arms and legs.

"What are they doing?" yelled Pauline excitedly from her workstation as she strained her neck to focus her mini camera on Fullview.

In an instant, the stalker raced forward, straddled the old man, and plunged the hand axe deep into his chest. Pauline and the rest of the members of the tribe were caught totally off guard.

"My god, they've killed the old man!" Both Pauline in her workstation and Pauline in Fullview simultaneously shouted. "They have killed my protector, my friend! They did it with *my* hand axe! What have I done to these poor erectus people? What have I done to him? He connected me with the erectus people—he saved me!"

Tears simultaneously flowed down the cheeks of both Pauline in her workstation and Pauline in Fullview.

"Pauline is transitioning back to her inner body, her homunculus," said Peachey.

"She is so distraught over what has happened she doesn't know what to do," added Vandervoort. "She is confused again, and her homunculus, the real Pauline inside, is coming back. It's coming back just like it did to differentiate the images we originally put into her brain from her own thoughts and just like it did when she had to confront the erectus women when they gave the predator call. The only way she can get out of these deadly situations is to let her homunculus take over completely, and I mean *completely!*"

"Those hands, those lips—my god!" exclaimed the paleoanthropologist. "I still can't get over that that is what we really are. That is our real power as human beings."

"Don't forget the tongue, don't forget the tongue!" yelled a smart-alecky Headley. "I wonder what the erectus people think of that tongue!"

"Can it, Headley!" yelled Peachey back over to Headley.

The transition of Pauline into her homunculus again startled the erectus people. They stepped back in awe. A few of the erectus children ran, but there was no predator call this time. Pauline's homunculus walked slowly toward the henchman who was still holding the hand axe. She thrust her huge right hand toward him.

"It looks as if Pauline wants him to give her the hand axe," I said. "Is she going to use it on him—I hope?"

The henchman threw down the hand axe as Pauline's homunculus approached close to him.

"What a chicken!" yelled Headley. "The big, tough guy is ready to run. It figures."

As two of the henchmen began to run, Pauline's homunculus transitioned back to Pauline's beautiful human body.

"Ah, she's back," Peachey sighed.

Pauline picked up the hand axe from where the henchman had thrown it and began to approach the stalker and his remaining two henchmen.

"Look, Pauline is holding the hand axe out to Mr. Pervert and his buddies," said Headley.

"His henchman has killed her friend, the old man, and apparently her inner body, her mammalian core, still wants to make peace," said Vandervoort.

"Yes," said Peachey. "Pauline's image-gathering electrode is in her mammalian brain, and this is where strong moral feelings first began to emerge two million years ago. She is simply doing what all 'normal' erectus people would want her to do, but beware, her stalker is not normal. As a child, he somehow missed the mammalian 'nipple-ization' experience."

With a sudden swipe of his powerful hand, the stalker sent the hand axe flying from Pauline's extended hand. As had been done with the old man, the stalker waved his remaining two henchmen to take Pauline to the ground.

"Oh, I can't look!" yelled Headley's girlfriend, Kimberly.

Headley gently put his hands on Kimberly's shoulders and turned her toward him. He then lightly tapped her on the end of her nose until her eyes opened so she could read his lips.

"Kimberly, no matter what happens, remember, remember, remember, this is all going on only in Pauline's mammalian brain," Headley reassured Kimberly. "Look over at Pauline in her workstation. See, she is okay. Although I see she is frowning about what is happening."

"You promise me she won't get hurt?" Kimberly asked.

"She won't get hurt, I promise," replied Headley as he continued to hold Kimberly by the shoulders while looking reassuringly into her eyes.

"But what is it like for her?" asked Kimberly. "Is it like a dream?"

"Well, you haven't been here at Fullview before, but Betty Jean can tell you she experiences everything we are watching," replied Headley. "At the same time, Kimberly, she can look out and see us watching her. For her it is like being an actor on a stage with an audience. But she has no script, and she doesn't know what the other 'actors' are going to do next. At the same time, Pauline is watching her 'deep-brain self' from over there in her workstation."

"Very good, Headley," I interjected. "It is kind of like Pauline is watching herself having a dream. But this dreaming is so real there will be many times when she momentarily gets caught up in the dream. Just like we all sometimes do. It's kind of schizophrenic, kind of not."

"Correct me if I am wrong, Betty Jean, but didn't you tell me the experience reminded you of Zhuangzi's parable where he told of dreaming he was a butterfly?" said Peachey.

"Yes, you are right, Dr. Peachey," I replied. "It is much like Zhuangzi dreaming he was a butterfly happily fluttering here and there then awakening in and feeling his human body. Zhuangzi said he wasn't sure whether he had dreamed of being a butterfly or the butterfly was dreaming it was Zhuangzi. It is a lot like that, because our inner body, our homunculus, is the butterfly. When your homunculus appears, you are dreaming. It, like Zhuangzi's butterfly, can flutter here and there. It can do anything it wants, it can solve any problem. But then you awaken to your outer body that has certain ... certain limitations, I guess, is the word."

"That's quite deep, Betty Jean," commented Vandervoort. "Would you say it was Zhuangzi's inner body that wrote the original parable of the butterfly?"

"Who else but the homunculus would know about such things as *being* a butterfly?" interrupted Peachey with a smile.

"All I want to know is who the hell this 'Zhawng Zoo' guy is," yelled Headley. "It sounds like he had an out-of-body experience to me! He was probably just high on something."

"Zhuangzi was an ancient Chinese philosopher, Headley," I quickly replied. "Over two thousand years ago, he wrote about the nature of transitions or transformations leading to higher levels of spirituality. Because I experienced transitions to my homunculus and back to my outer body when I was in Fullview, I became interested in his philosophy. I became interested in what it all meant."

"By the way, everyone," added Peachey. "Zhuangzi probably did not have an out-of-body experience per se, but it is interesting that you brought that up. We now have electrode evidence that, in fact, it is the homunculus that has the out-of-body experience. The homunculus, which sees the *outer body* in the mirror many times every day, tries to grasp onto its outer body when it thinks the outer body might be leaving it for good, like when it thinks it is dying. Every dream we have may actually be a mini out-of-the-outer-body experience. Since our homunculus can detach itself and imagine flying above the familiar image it sees in the mirror, it can also watch what we are doing when we dream. I think I can rest my case on that!"

"God, look at them!" exclaimed the paleoanthropologist. "They are gnawing at her body!"

The two henchmen spread Pauline's naked body out on the ground and commenced gnawing at it. At the same time, they waved the stalker to join in. He had other ideas. The stalker had retrieved the hand axe and was standing over Pauline with it raised high above his head.

"It appears the stalker is about to perform a *ritualistic* killing of Pauline," said the paleoanthropologist.

Suddenly, Pauline transitioned to her homunculus form, her huge hands brushing aside the two henchmen.

"Look!" exclaimed one of the female technicians. "The head of that good-looking tall erectus man is transitioning. Where did he come from? I hadn't noticed him before."

"What the hell?" said an astonished Headley.

"I can't believe what I am seeing!" said Pauline from her workstation, where she had turned to focus her mini camera on Fullview.

"By god!" said the paleoanthropologist. "It's Headley! The handsome erectus man is becoming Headley!"

Everyone stared at the erectus man's head, as it slowly became Headley's head.

"Jay Headley!" yelled Headley's Kimberly. "What are you *doing* in there?"

Headley began to turn a brilliant *"embarrassment red"* as he watched a duplicate of his head being erected on the body of the tall erectus man.

"Hey, that looks more like Headley than Headley does," commented Peachey.

"But just what exactly are you doing in there?" said Headley's Kimberly as she looked jealously into the eyes of a red-faced Headley.

"I'm not in there," replied Headley. "I mean, that's not me in there. Help me, Dr. Peachey. What's going on?"

"That erectus man with Headley's head is Pauline's brain mate," replied Peachey. "Remember when Betty Jean had been devoured by her dinosaurs? As a last resort to stay alive, her brain called forth Larry, her brain mate, to rescue her."

"That's right," said Vandervoort. "Pauline has no current mate, but her brain has chosen Headley to try to save her."

"Why Headley, of all people?" Peachey said with a snide tone.

"I think she has chosen Headley because he has been with her for years in the lab," replied Vandervoort. "Pauline trusts Headley. Because he's brash and cocky, she sees him as dominant and resourceful. She wouldn't want a dumb old professor like me in there, trying to save her!"

"By the way, Kimberly," said Peachey. "From one woman to another, that's not your Jay Headley in there. Jay has nothing to do with it. This is all going on only in Pauline's brain. She is probably as surprised as you are to see Headley in Fullview. This animus-anima brain mate stuff is all going on at an unconscious level in Pauline's brain."

"Amen to that," said Pauline as she looked on from her workstation. "But, I must admit, I am damn glad to see Headley in there with me right now."

"Pauline's homunculus is standing up, confronting the stalker," said Headley's Kimberly. "Her homunculus is mouthing the erectus predator call right in his face."

The erectus stalker was momentarily frightened by the sudden return of Pauline's homunculus. The predator call further unnerved him. As he stood momentarily stupefied, Pauline's erectus-Headley brain mate leaped toward the stalker.

"Pauline's Headley-incarnated brain mate just snatched the hand axe from the stalker!" proudly exclaimed a cheerleading Headley.

"Go get 'em, Headley," Vandervoort chimed in.

The erectus stalker picked up one of the heavy hammer stones that the erectus woman had been using. Pauline's Headley-erectus brain mate and the stalker squared off for a battle.

"Why is Pauline's brain mate throwing away the hand axe?" I asked. "He's throwing away his advantage!"

Pauline's Headley-incarnated brain mate began to maneuver like a man skilled in martial arts.

"Ha!" exclaimed Headley from the other side of Fullview. "He's not only got my face, he's got my brain. He knows the same martial arts I know. He doesn't need that stupid hand axe!"

"Arrogant to the end," Peachey sighed.

Pauline's brain mate quickly executed several fast, hard-hitting martial arts blows to the erectus stalker's head and body.

"Boy, that was quick!" said one of the female technicians.

The erectus stalker lay moaning in a crumpled heap. The rest of the erectus people, including the stalker's henchmen, stood in silent confusion and awe at the sight of the transformed Headley-erectus and his strange movements and skills.

"Pauline is transitioning back to her beautiful human body," yelled Kimberly from the other side of Fullview.

"My god, what is she doing?" said the paleoanthropologist. "After all that, she is going to help the stalker?"

Pauline walked over to the erectus stalker, who was badly beaten and nearly unconscious. She knelt beside him, caressing his head and humming softly to him. She then stood, embracing erectus-Headley and thanking him in her plain English, a complex type of utterances the erectus people had never heard but erectus-Headley understood.

"Look over at Pauline in her workstation," said Peachey. "She is crying. Why is she crying, Dr. Vandervoort?"

"Excuse me, Dr. Vandervoort," I interrupted. "I know why she is crying, Dr. Peachey. She has met her brain mate and is realizing, as I did, that she has never really been alone, will never be alone again. You suddenly realize in a gestalt flip that the perfect mate you have been looking for has been with you all the time. And you cry. You can't control it."

"Thank you, Betty Jean, you said it better than I could have. You were there," added Vandervoort. "We think the brain mate developed in the early human brain with *Homo habilis* and *Homo erectus*. It is part of the caring, nurturing, and long caretaking that human children require. It is part and parcel to being a mammal, it is part and parcel to the nipple-ization of *Homo erectus*."

"But why the *reflexive* crying?" asked a still-tearful Pauline. "It is so overwhelming!"

"It's pretty simple," interjected Peachey. "You and Betty Jean have been the only ones to actually see their brain mates and to realize that they are actually part of your brain. The crying is happy crying, it comes from the sudden jolt as you have suddenly climbed out of the deep darkness of the dinosaur brain and into the emotion and brain mates of the mammalian world. These brain mates that have appeared in Fullview are solid proof that the ancient brain tissue and the ancient images are the same thing. Once you have an erectus mammalian brain, you have a living brain mate forever! A living, holographic brain mate that is as alive as any part of you."

"That's right, Pauline," added Vandervoort. "Just as sex and aggression are actual parts of the dinosaur brain in us, our brain mates are actual parts of our mammalian brain. They are there to inform us as to what to look for in a mate, to help us take care of the young, and to tell us to protect and take care of each other. From now on, the job of survival requires the two of you. Just as it was for *Homo habilis* and *Homo erectus*, you have been equipped to survive best by the real or the imagined presence of your brain mate. Dinosaurs could never be happy, they could never cry. You can—you can because you care."

Everyone was silent. All activity inside the huge Fullview monitor had stopped. Both Pauline in her workstation and Pauline in Fullview were completely still. All eyes were pensive.

"Does this mean that my Headley is Pauline's 'forever' brain mate?" asked Headley's Kimberly, breaking the silence.

"I'll let you answer that, Dr. Vandervoort," said Peachey.

"Maybe, Kimberly, but probably not," said Vandervoort. "I think the fact that Pauline's brain materialized only Headley's head and some of his skills is very meaningful."

"What do you mean?" asked Kimberly.

"Pauline's brain materialized only the parts of Headley she needed to take care of the situation at hand," replied Vandervoort. "She knew Headley was her friend and that he is a cocky, tough sort of a guy who had studied martial arts. That's what she needed to take care of the stalker."

"Yes," added Peachey. "If she had been romantically involved with Headley, she wouldn't have put his head on the erectus man's body. She would have put it on a *Homo sapiens's* body, a body like Headley's real body."

"What they are saying is true," said Pauline from the workstation and from Fullview simultaneously. "I love you, Jay Headley, but not in a romantic way. I love you like a brother. That's the way you have always been to me."

"I understand, Pauline," said Headley as he approached Pauline's workstation with Kimberly. "That is the same way I have felt about you during these years we have worked in the lab together. You are my little 'sister,' Pauline, and you always will be."

"Hi, sis," said Kimberly as she smiled and reached out her hand to Pauline.

"Something about all of this has been bothering me," I said. "I am not sure what it is or even how to talk about it."

"Well, try me," said Vandervoort and Peachey almost simultaneously.

"Okay, you guys," I continued with a smile. "I keep thinking about my journey to my land of the dinosaurs. It wasn't *the* land of the dinosaurs. It was *my* land of the dinosaurs."

"What's the difference?" said a slightly puzzled Headley. "*The* land, *my* land, what's your point?"

"Well, since it was *my* land of the dinosaurs, everything that went on in Fullview was really only about me," I continued. "It was, wasn't it?"

"No, Betty Jean, it wasn't just about you," replied Vandervoort. "Your journey to the land of the dinosaurs was everybody's journey. Remember, these journeys are coming from tens of millions of years of images that all people have in their brains. And the brain mates have been in the advanced mammalian brains of everyone for the last two or three million years, maybe eight million years. So brain mates exist in everyone's brain. Just as they did for Pauline, they would show up as *perfect* brain mates in Fullview."

"Betty Jean, are you troubled by the way Pauline was able to put Headley into the erectus tribe to help her?" added Peachey. "Maybe that made it seem it was all about Pauline, and that she had control over what happened. Is that what is bothering you?"

"Yes, that is kind of what is bothering me," replied Betty Jean.

"Ah, yes," said Vandervoort. "I see what you mean, but remember, Betty Jean, a small part of your images in Fullview come from your own recent life experiences."

"Sure, Betty Jean, you remember me mentioning Penfield's findings that in each of our brains there is a permanent record of everything, every image we have experienced," added Peachey. "That's how Headley got into Fullview. He came right out of Pauline's personal permanent record of memories."

"But did Pauline have actual control over getting Headley into Fullview to help her?" asked Headley's Kimberly.

"I don't think she had *control*," replied Vandervoort. "The brain has all these millions of years of images, images right up to the present. It keeps them all in memory and uses them with a constantly sharpening efficiency to solve any kind of problem that might come up. The brain uses them silently from its unconscious memory archives. Pauline's brain created the erectus people because it already had those images. Her brain created the old man, the stalker, the predator calls, the hand axe—all of it—because it had those images organized in layers in and surrounding her mammalian brain."

"Those millions of images streamed into Fullview in a way that made sense to the unconscious processes of the brain," added Peachey. "We have been watching how the living brain thinks! Most of it goes on outside of the direct control of the immediate twenty or thirty seconds of conscious awareness of the person."

"*But*" said Vandervoort loudly, pausing to get everyone's attention, "the great value of Fullview is to actually see all of this unconscious stuff going on so that we can know how our living thought process goes on. It has two great values. Number one, we are recording these image algorithms or methods of thought so that we can build Nanosoft's Living Windows. You all know that. But what you may not know about is the second value. Once Betty Jean and Pauline saw the origins of their feelings and thoughts and watched their brains deal with them, they were no longer afraid of anything. Pauline now knows that she has millions upon millions of images in her brain that she can use anytime she wishes."

"'She has one hundred thousand generations of seeing and feeling things she can use anytime she wishes," said a reflective Pauline as she sat in her workstation, staring at her beautiful self in Fullview. "There is a normal visual world of dinosaurs, erectus people, and Neanderthals waiting for me to meet them in Fullview."

As Pauline wiped a tear from her eye and relaxed, activity in Fullview suddenly resumed. The erectus people gathered around Pauline, erectus-Headley, and the stalker and his henchmen. The old man who had been killed by the stalker lay in the middle of it all, as if he were sleeping.

"Look," said the paleoanthropologist, "Erectus-Headley's head is transitioning back into the erectus man's head."

The stalker and his henchmen rose to their feet.

"It's like they are all actors just completing a play," said Peachey. "Each of the actors is taking their place as if they are all going to take bows."

"It does look like that," said Vandervoort. "But I think what we are seeing is a change in Pauline's entire understanding of the situations going on in her own mind. She now knows that these people, the roles they each play, and these situations have all been assembled by what the brains of her ancestors had repeatedly experienced over millions of years."

"And also by who she is in her unique contribution to the All Time," added Peachey.

"So what has happened to her with the erectus people has taught her that she can be in control of what happens?" I asked.

"In a way, yes, Betty Jean," replied Peachey. "She has learned that she can be in control in Fullview if she really tries. If anyone really tries, their brain will use every resource, every image that it has to take care of you."

"That's what it was like for me in Fullview," I added. "I was even eaten alive by my dinosaurs. God, that was nice! But my own brain rescued me, and I learned I could control my dinosaurs in any way I wished. If I went back to my land of the dinosaurs in Fullview, I would let them eat me again, and I would let my brain mate rescue me again, and again, and again!"

"Easy there, Betty Jean," said Vandervoort with a smile.

"Hey," yelled Headley. "The erectus people are taking turns hugging the stalker, and they are hugging his henchmen too."

"Whoa!" exclaimed Headley's Kimberly. "The old man's eyes are opening."

The old man rose to his feet and looked around at everyone hugging and smiling.

"He looks puzzled but happy," said Vandervoort. "He doesn't even know he'd been killed."

"But I do," said Pauline from inside Fullview. "The old man will go on and continue to be the wise man in the mammalian part of my brain. He is my friend forever. And now he knows it!"

As Pauline in her workstation watched and listened to Pauline in Fullview, a tear ran down both their cheeks.

"This is absolutely amazing, Professor Vandervoort," said Jennifer, one of the female technicians. "Pauline in her workstation is feeling right along with Pauline in Fullview."

"Yes it is, Jennifer," said Vandervoort. "Remember, Pauline is experiencing her mammalian brain, which is the feeling and *believing* emotional brain. Both she and her inner body or homunculus inside Fullview are feeling the same emotions at the same time."

"Pauline is as wise as the old man who lives in the ancient part of her brain," said Peachey. "Before Pauline leaves the erectus people, she wants everything to be happy there. It is, after all, Pauline's world that we are all witnessing."

"Dr. Peachey," said Pauline from her workstation. "Could I stay here with the erectus people if I wanted to?"

Peachey looked at Vandervoort, somewhat caught off guard and dismayed.

"Well, that's an interesting thought," Peachey replied. "Why do you ask?"

"The people are so simple and so caring," replied Pauline. "I really don't want to leave yet. I could teach them so much, and I would like to explore their world."

"Dr. Vandervoort, how long could we let her stay with the erectus people?" asked Peachey.

"Would you like to stay another hour or two, Pauline?" asked Vandervoort. "That should be enough time to do quite a lot."

"Only an hour or two?" replied Pauline.

"An hour or two would give you a lot more time with the erectus people than you might think, Pauline," replied Vandervoort. "Look at all of the things you have already experienced with them in just a few minutes, nine minutes to be exact. You got their confidence by making their predator call, you made them a hand axe, the old man was killed, you defeated the stalker, and brought the old man back to life in exactly nine minutes!"

"I have only been with the erectus people for nine minutes?" exclaimed Pauline in disbelief.

"The actual time here in the Fullview lab and the kind of time you spend with the erectus people two million years ago don't run at the same rate," interjected Peachey. "When you are with the erectus people, you are experiencing thoughts and feelings at the same speed you can think or dream. You could have a whole life with the erectus people while only a couple of hours would go by for the rest of us in the Fullview lab. Remember, when you are with the erectus people, you are in Fullview's holographic nonspace in All Time. In nonspace, time is greatly distorted, even infinitely distorted."

"Tell her about the Peachey Time Sled, Dr. Peachey!" I excitedly said.

"Hmmm," said Peachey, raising an eyebrow at me. "Just let me say, Pauline, that under the right conditions, you can go to your land of dinosaurs and back at the speed of light, and it *feels* like a sort of timeless moment. It is a pure 'holo-moment' containing millions of images spread across millions of years, and that underlies and gives meaning to your everyday experience! In other words, a ride on the Peachey Time Sled is the speed-of-light summation that you have indeed been alive for millions of years and that you are alive right now!"

"What Betty Jean and Dr. Peachey are alluding to is the fact that time is always very distorted in nonspace because you are traveling in the All Time," added Vandervoort. "The All Time is the time that lives in the millions upon millions of images permanently stored in the layers of your brain. There is no tick-tock in the All Time of nonspace. The *time* stored in your brain can run at any speed the overall circumstances require."

"And I will think that speed is *normal*?" asked Pauline.

"You will think that what you are experiencing is totally normal," replied Vandervoort. "Remember, Einstein showed us that no matter where you are, the speed of light is the speed of light, and that makes wherever we are, no matter how fast we are traveling, and whatever we experience seem normal to us."

"I could live a thousand lives in nonspace while, at the same time, you guys watching me in Fullview would live only one life, right?" asked Pauline.

"That is correct," said Vandervoort as he looked over at Peachey as if a light had gone on.

"Would I ever have to die in the All Time of nonspace?" asked Pauline. "I mean, if I could live a thousand lives while you guys live one, why can't I live a million, why not a trillion?"

The room was silent for a moment. Everyone waited for a response from Professor Vandervoort.

"Pauline, there is no reason you couldn't live a million lives in Fullview," replied Vandervoort. "Would you ever have to die in the All Time of nonspace? We just don't know the answer to that intriguing question. But remember, Pauline, it *is* the All Time."

"Dr. Vandervoort, I want to stay with the erectus people," responded Pauline. "I want to help them, and I want to take a mate and have children here. I want to stay for at least twenty years in the nonspace of two million years ago. Can I do that?"

"You can," said Vandervoort. "And we will be with you, Pauline. We will be watching."

By early that same afternoon, we had watched Pauline's life among the erectus people unfold. It had literally unfolded from the nonspace record stored in her mammalian brain. Over the twenty years she experienced, she had had three children and had become the leader of the erectus women. The women had made thousands of hand axes, and there had been no more *stalkers* and no more *henchmen*. The tribe had an abundance of food, and the proliferating young were learning new things quickly.

At 2:05 p.m., Pauline suddenly arose from her workstation. According to our clock, we had been watching Pauline live a twenty-year *lifetime* in about three hours. We were shocked at her first words.

"Could I have some coffee and something to eat?" said Pauline. "If I don't eat something, I think I will collapse!"

The three hours we watched twenty years of Pauline's life among the erectus people was like watching a movie. But this movie had a fine detail of overtones and recurrent patterns of thought that somehow told you that it was all real and somehow all familiar. At the same time, in this movie, we never knew what was going to happen next and neither did any of the people in the movie. I guess this is why we have plays and movies, and I guess that is, in fact, what they are—they are journeys in our own brains.

"Someday I will go back to visit my erectus people," Pauline suddenly said. "I wonder if they will remember me. I wonder if my children will remember me. God, I wonder if they will miss me."

"If you go back, Pauline, they will remember you," said Peachey. "They are now part of your permanent record of memory. And don't worry, they won't miss you. Until you go back, they are simply asleep in the records of your mammalian brain."

"Will they age while I am gone?" asked Pauline.

"No, Pauline, they won't age one minute," replied Vandervoort. "You won't have aged either. Your body and mind are waiting for you just as you were two million years ago back in the nonspace of your mammalian brain. You can pick up where you left off and go back and forth a million times, and you will all slowly age by how much nonspace time you spend there. No one, not even *you* most of the time, will even know you have been in and out of mammalian nonspace."

"Wow!" yelled Headley. "That's cooool!"

Pauline finished her small lunch and walked back to her workstation.

"I am going to initiate my Journey 1 electrode, Betty Jean," said Pauline. "I am going back one hundred million years to my land of the dinosaurs. I want to have the experience you had."

"I'll be watching you in Fullview," I replied with a knowing smile. "Ah, mmm, go onward to the joy of being eaten!"

"Even if she stays a million years, she'll be back in thirty minutes," said Peachey. "She will be very much refreshed, very much!"

"Yes," said Vandervoort. "In Fullview, Betty Jean and Pauline have shown us that the All Time is with us all for all time. The future is bright for the erectus people. The future is bright for all of us!"

Colonel Edythe Peachey (Retired)
December 24, 2014

AFTERWORD

In the year that followed, Pauline continued to explore the All Time. There were two hundred million years of archived experience awaiting her. In the All Time, she was experiencing a lifetime at a time. We began to discover that the precise moment she happened to choose to enter the All Time determined the particular lifetime thread that came to life. This precise moment has the same effect for her as how the precise moment you happen to get into your car to drive to work or to the store determines the traffic you will encounter and the people you will see. There were approximately three million lifetimes for her to live. But on top of that, there were literally billions of different entry points or precise moments for entering her holographic brain.

And we were coming to the realization that whenever Pauline was in Fullview, her existence was electronic, and for that time, she was ageless. We had duplicated her entire being in the form of the holocopies of Fullview. For all practical purposes, Pauline was immortal and could live in any of the countless lives layered into her two hundred million years for as long as she wished. And now, anyone could follow Betty Jean and Pauline forever into the All Time. We also came to realize that all of us had already lived the two hundred million years laid down in our human brains. This life we live is one long life that began, not when we were infants, but began when we were forming the first layers of our brains two hundred million years ago—that has been our real life. This entire ancient life force within us naturally strives to continue seeking immortality as it has all along. And with the human brain, those two hundred million years of struggle became aware of this immortality, and this brain has been given the ability to achieve it.

As the press slowly began to find out more about Fullview and the All Time, I had to slowly reveal more and more about how it all worked. To make a long story short, Pauline's beautiful black homunculus made the front page of newspapers all over the world. I told the world that we had found that the talking box in each of us that originally allowed us to escape the constraints of space and time millennia ago was also our way out of the body and into the eternity of the All Time.

Based on holocopies of Pauline's three layers of brain algorithms, *Living Windows* 2016 is now in its final stages of development. My understanding is that it will be released on Monday, January 4th, 2016.

Professor Vandervoort predicted that by the year 2044, the choice to live in Fullview eternally (the Forever Time) would likely become as commonplace as owning an iPhone in 2014. We did know that the 2044 Project was beginning to recognize that the entirety of a person's personal All Time was the only way into their Forever Time. That is, the 2044 Project was finding, as we did, every individual must take their own dinosaurs, own *Homo erectus* people and own brainmates as they take up residence in Fullview.

Colonel Edythe Peachey
Currently back on the *Ti Amo*, Pier 39, San Francisco

Other Published Works by Lawrence Vandervert

"Understanding Tomorrow's Mind: Advances in Chaos Theory, Quantum Theory, and Consciousness in Psychology," The Journal of Mind and Behavior (1997) [editor, special double issue];

"How Working Memory and the Cerebellum Collaborate to Produce Innovation and Creativity," Creativity Research Journal (2007);

"Cognitive Functions of the Cerebellum Explain how Ericsson's Deliberate Practice Produces Giftedness," High Ability Studies (2007);

"The Emergence of the Child Prodigy 10,000 Years Ago: An Evolutionary and Developmental Explanation," The Journal of Mind and Behavior (2009);

"Working Memory, the Cognitive Functions of the Cerebellum and the Child Prodigy," Springer Science (2009);

"The Evolution of Language: The Cerebro-Cerebellar Blending of Visual-Spatial Working Memory With Vocalizations," The Journal of Mind and Behavior (2011);

"How the Cerebro-Cerebellar Blending of Visual-Spatial Working Memory with Vocalizations Supports Leiner, Leiner and Dow's Explanation of the Evolution of Thought and Language," The Cerebellum (2013);

"Working Memory in Musical Prodigies: A 10,000 Year-Old Story, One Million Years in the Making" (in press), Child Prodigies in Music, Oxford University Press;

"The Collaboration of the Cerebellum and the Cerebral Cortex in Working Memory: A Case Analysis of a Child Prodigy" (in press), Child Prodigies in Music, Oxford University Press.

See additional research publications by Lawrence Vandervert on the Internet at Larry Vandervert, ResearchGate.

www.ingramcontent.com/pod-product-compliance
Lightning Source LLC
Chambersburg PA
CBHW031055180526
45163CB00002BA/843